Sir Benjamin Baker

Long-span railway bridges: Comprising investigations

With numerous formulae and tables

Sir Benjamin Baker

Long-span railway bridges: Comprising investigations
With numerous formulae and tables

ISBN/EAN: 9783744726962

Printed in Europe, USA, Canada, Australia, Japan

Cover: Foto ©berggeist007 / pixelio.de

More available books at **www.hansebooks.com**

LONG-SPAN RAILWAY BRIDGES;

COMPRISING

INVESTIGATIONS OF THE COMPARATIVE, THEORETICAL AND PRACTICAL ADVANTAGES OF THE VARIOUS ADOPTED OR PROPOSED TYPE SYSTEMS OF CONSTRUCTION,

WITH

NUMEROUS FORMULÆ AND TABLES,

GIVING

THE WEIGHT OF IRON OR STEEL REQUIRED IN BRIDGES FROM 300 FEET TO THE LIMITING SPANS.

BY

B. BAKER.

ILLUSTRATED BY PLATES.

REPRINTED FROM ENGINEERING.
THE WHOLE CAREFULLY REVISED AND EXTENDED.

PHILADELPHIA
HENRY CAREY BAIRD,
INDUSTRIAL PUBLISHER,
406 Walnut Street.
1870.

PREFACE.

The contents of the following pages have already appeared in the columns of *Engineering*. The purpose of this "*replica*" is to present the revised results, and tables, in a more accessible form than they could attain scattered in a desultory manner through successive numbers of a periodical.

The subject, in its present form, was suggested by the discussion, at the Institution of Civil Engineers, following Mr. Barlow's paper on the Clifton Suspension Bridge; when the absence of any simple generalization of the question was evidenced. Prior to that time, however, the consideration of "Long-span Railway Bridges" devolved upon the author in the course of his professional duties, and some valuable data had accumulated. On proceeding with the investigation, it was at once seen that a strictly mathematical treatment of the subject would entail lengthy and involved formulæ, and absorb far greater space than was available for the purpose; indeed, the works of Gaudard and Schwedler, treating on the same subject, but within very narrow limits, plainly illustrated this fact. Accordingly, the various hypotheses, which it is absolutely necessary to make in an inquiry of this nature, are framed as comprehensively as possible; and in many instances the result of a careful balancing of probabilities is given without exhibiting the process by which it has been evolved. In short, elimination, and not elaboration, has been the aim throughout.

B. B.

CONTENTS.

                                                                                                           PAGE

Definition of long-span bridge—Circumstances justifying adoption—General problem of indefinite solution—Importance of superstructure—Want of information on the subject—Possible to approximate—Object of paper, and method proposed to be adopted . . . . . 9—13

Relative strains on webs and flange—Lattice and plate webs—Minimum thickness of plate governs depth—Limit of depth for lattice—Method for arriving at weight of ironwork necessary—Limiting strain—Types of construction to be investigated in iron and steel . . . 13—16

Type 1. Box girders—Economic depth small—Buckling of plates and waste of metal at centre—Weight of web governs depth—General principles determining proper depth—Formulæ for depth, and minimum thickness of plate allowable—General formulæ for strains and multiples—Tables of same . . . . 16—21

Theoretical deductions overruled by practical—Position of plate and lattice reversed—Advantage of lattice on account of increased economic depth—Stiffness—American and English lattice—Investigation of strains, and proper depths of Type 2—Lattice girders—Tables of strains and multiples . . . . 21—25

Type 3. Bowstring girders—Uniform and unequal loads—Bracing requires little metal, otherwise proves disadvantageous—Stiffness—Investigations of mass required.—Influence of load—Strain on diagonals—Tables of strains and multiples . . . . 25—32

	PAGE
Type 4. Straight-link girders—No example in this country—Chepstow bridge—Bollman truss—Discussion at Institution—Similarity to bowstring—Influence of load—Reversed conditions—Inclination of bows—Deflections peculiar—Practical difficulties—Strains, masses, and multiples—Tables—Conclusion of first stage	32—41
Type 5. Cantilever—Appearance same as Type 2—Web same also—Mass and moment of flange theoretically less—Extra metal in adjacent spans diminish economy—Mass of metal—Tables of constants, strains, and multiples	41—45
Type 6. Cantilever varying depth—Modification of form desirable—Curved-top flanges diminish strain on web—Depth and stiffness greater—Mr. Fowler's Severn bridge—Tables of multiples, &c.	45—48
Type 7. Continuous—Maximum span—Sedley's bridge no practical difficulty—Economical results surpass other systems—Tables	48—53
Type 8. Arched rib—Early form of long-span bridge—Indefiniteness of opinions concerning it—Hypotheses vary—Rib jointed at three points—Nicety in proportioning metal to strain—Expansion and contraction entail no additional strain—Investigation of strains—Tables, &c.	53—60
Type 9. Suspension with stiffening girder—Lightness and strength of rope between supports—Earliest form of long span—Instability at first inadequately appreciated—Check ties—No important difficulty in obtaining any required amount of stability—Transverse strength necessary—Vibration always exists, but not detrimental—Great variety of designs for rigid bridges—Investigation of strains on chains and stiffening girders—Tables, &c.	60—66
Type 10. Suspended girder—Reasons for rigidity of girder as compared with suspension—Inverted bowstring—Reduction of mass by initial tension on boom—Mr. Fowler's Thames bridge—Freedom from vibratory impulses—Expansion provisions—Investigation of strains, masses, &c.—Tables	66—71

PREFACE. vii

	PAGE
Type 11. Same as straight-link girdle, less the boom—Rolling—Dead and mixed loads—Tables of strains and multiples.	71—73
Conclusion of tables of multiples—Investigation of specific loads—Weights of platform—Cross girders—Useful load—Formulæ and classification—Gross loads	73—77
Tables of loads and weights of iron in cwts. per foot run	77—81
Adaptation of formulæ to steel structures—Available strength for useful load higher ratio than limiting strains—Tables of multiples, strains and weight of metal	81—91
Summary of results, illustrated by diagrams—Comparative economy refers to superstructure only—Expense of piers influences economy—Fairest comparison obtained by taking average weight per foot of viaduct—Tables of same—Difference less marked—Influences of cost of scaffolding on designs for long-span bridges	91—97
INDEX	99-100

LONG-SPAN RAILWAY BRIDGES.

According to Dr. Johnson, a bridge is "a structure to carry a road across a watercourse;" and although this interpretation of the word is not sufficiently comprehensive to include all cases in the present progressed stage of the art of building, yet, if we limit its application to *long* spans alone, we may even render it still more explicit, and with very little liability to error define a long-span bridge to be a structure for carrying a *railway* across a watercourse. The reasons why this is the case are sufficiently obvious; in the first place, the condition necessitating the adoption of a long span is generally either that a certain width of opening must be provided clear of all obstructions, or that the expense of carrying up a number of lofty piers is, owing to some difficulty in obtaining secure foundations, so great as to render it more economical to reduce the number of individual supports, and so concentrate the resulting greater load on fewer points. Neither of these conditions is likely to occur, except when a watercourse is the obstacle to be surmounted, when, probably, a navigable channel of certain width has to be maintained, with sufficient headway to admit of the free passage of vessels. If in any case it should be desirable for the span to be greater than the minimum amount dictated by the compliance with these conditions, it could only be when the depth and rapidity of the current, or the treacherous nature of the bottom, rendered it desirable to reduce the risk of construction to a minimum. As steep gradients are now worked with ease and economy, it is not at all probable that any other case —such as that of carrying a line across a ravine—will ever occur in which it would be economical to introduce longer spans than 300 ft.; and it is only spans above that amount we designate *long* spans in the present paper. Again, it is highly improbable that any long-span bridge should be other than a railway bridge, because the great expense involved in the construction must be justified by necessity; in other words, by the probability of such large traffic as a railway alone could accommodate. Even when we have thus limited the question to railway bridges crossing a watercourse, where a given span and height have to be maintained, we have by no means obtained all the conditions enabling us to pronounce upon the proper type of construction to be adopted. Thus, if the banks of a river are lofty, and afford a firm foundation for the superstructure with little or no piers, an arch or suspension-bridge may possibly be the most economical construction, although the resulting span may be greater than absolutely required; and if the banks are nearly level with the stream, it may or it may not be advisable to make the adjacent spans of greater length and weight than would ordinarily be required, in order to enable them to contribute more effectively to the support of the larger centre span. In short, it is plain that the determination of the most economical construction for a bridge of given span is a problem admitting of no definite solution. We may, however, facilitate the process much, and obtain valuable positive results, if we confine our attention at first to the comparative weights of iron required in the different methods of constructing the superstructure, which, after all, is by far the most important element in determining the cost of a long-span bridge.

The size of a bridge is very commonly the popular standard by which the eminence of its engineer is measured; we may, therefore, naturally expect to find engineers ambitious of excelling one another in this particular branch of their profession, but for the same reason, as so much consideration must necessarily at one time and another have been devoted to the elucidation of the subject, a student of engineering may justly be surprised to find so little definite information existing as to the capabilities of all possible combinations of design to do the required work effectively and economically. Yet the number of patents taken out by professional and other quacks indicates clearly the want of appreciation of the fact that the problem is one admitting of a rigid theoretical solution, and that the limit beyond which the quantity of metal required in the actual construction exceeds the amount theoretically required, will be a factor of the latter quantity, the value of which may be approximated to very nearly, if we avail ourselves of the stock of information afforded by existing though similar structures.

Although, as we urge, it is possible to approximate very nearly to a true result in every case, a considerable amount of intricate calculation and considerable space would be required to treat the subject exhaustively. Yet a great deal may be done with little labor if we base our investigations on the simplest and broadest principles, avoiding all complications and neglecting altogether minute details. If we proceed thus, we may, by making, so to speak, a cartoon of the different systems, exhibit in bold outline the respective advantages and disadvantages appertaining to each. This will be our aim in the present paper. We shall investigate on the above broad principles the weights of different types of girders, including all probable combinations, from the minimum span of 300 ft. to the limiting span, beyond which it would be impossible to construct a bridge of the class capable of carrying more than its own weight without exceeding the given limiting strain per square inch. We shall carry out similar investigations both for iron and steel, and so conduct them, that by arranging the results graphically in the form of a diagram we may obtain a comprehensive view of the properties of the different designs and the nature of the laws governing the increase of weight, and consequent relative cost of the different constructions in the two materials.

. The general principles on which we shall proceed are identical with those already advanced by the author in a paper on "The Proper Depth of Girders" (*Engineering*, vol. ii. p. 224). We shall, where we consider it advisable, sacrifice mathematical exactness of formulæ to simplicity, and generally allow a very free interpretation of theoretical deductions; the numerical results will be worked out with the slide rule, and, in short, the process throughout will be consistent with our professed object of exhibiting a cartoon-like view of the subject under consideration.

In the paper alluded to, we observed that the maximum strain on each flange of a girder of uniform depth is, by the simple principles of leverage, equal to the distributed load multiplied by the span, and divided by eight times the depth, the strain being greatest at the centre, and less elsewhere in proportion to the ordinates of a parabola; and as this strain on the flange could only have been transmitted through the web, a little consideration will make it evident that for a distributed load, whatever the resultant strain may be, the total amount on the half web, resolved horizontally, must be equal to the maximum strain on the flange, or, in other words, the horizontal sectional area of a plate web should be equal to double the sectional area of the flange at the centre. For a lattice web the gross sectional area of all the bars should be greater than this in the same ratio as the length of a bar exceeds the horizontal distance included between its two ends, which for the most economical angle of 45° for the bars amounts to $\sqrt{2}$ times the section required in a plate web, and the mass will consequently be $\sqrt{2} \times \sqrt{2} =$ double that of the plate web. Expressed algebraically, S being the span, and d the depth in feet, a the sectional area of *each* flange, x and y coefficients depending upon the practical construction of the flange and web respectively; the mass of the plate girder will be $=2a\,(Sx+yd)$, and that of the lattice girder will be $=2a\,(Sx+2yd)$.

With the plate web, we found the variable ratio of d to S to be governed, chiefly, by the minimum thickness of plate allowed in the construction of the web, which we considered should increase with additional loads per foot on the girder. With the lattice girder, we considered that, whilst the weight of the flange would in all cases be inversely proportional to the depth, that of the web, beyond certain limits, would increase directly as the depth; that is, although theoretically constant, there would be some limit beyond which the additional strength required in the struts as columns would necessitate a greater quantity of metal than the shorter struts with the heavier direct compression. This limit we have in all instances considered to be reached when the weight of the web becomes equal to that of the flanges. That is, since the mass of the flanges, $2Sx$, increases inversely, and that of the web, $4yd$, increases directly as the depth, the minimum value of the whole mass $2Sx+4yd$, will occur where $2Sx = 4yd$. The effect of this in the different constructions will be controlled and adjusted for each case by the variable values of the coefficients x and y.

In order to arrive at the weight of iron required in the construction of a girder capable of carrying a given load with a given maximum strain per square inch, we shall find it convenient to invert somewhat the ordinary method. Thus, instead of starting with the load and determining the strain on that data, we shall deduce the former element from the latter. Taking 1 square inch area of flange as the section resisting the maximum horizontal strain on that member, the mass of the lattice girder will be $2(Sx+2yd)$; and from that we can easily determine the moment and the strain on the flange due to the load of the girder itself, which would obviously be the same amount per unit of area, whatever the gross section of the flange might be. We have given, the limiting strain per square inch due to the entire load $= T$, the strain per square inch due to the weight of the girder itself $= t$, and, consequently, we have also $T - t$, the strain available for the useful load carried. Again, it follows that—moment of weight of girder : moment of useful load :: $t : T - t$; and that, if the weight of the girder be uniformly distributed in the same manner as the remaining load, we have—weight of girder : useful load :: $t : T - t$. Or, in other words, the weight of iron required in the construction of a girder to carry a given load will be the multiple $\frac{t}{T-t}$ of that load.

By adopting the above method, we have much facilitated the solution of the problems before us, as we can now proceed with our investigations without complicating the question by the introduction of the varying loads, to which we are in each individual case liable.

The types of construction to one or the other of

which we consider all forms of bridges, not absolute eccentricities, may be referred, and to which, consequently, we have confined our investigations, are the following:—

1. Box-plate girders, including tubular bridges.
2. Lattice   do.   do.   Warren truss, &c.
3. Bowstring do.   do.   Saltash type.
4. Straight links and boom. Bollman truss.
5. Cantilever lattice parallel depth.
6. Do.   do. varying economic depth.
7. Continuous do.   do.   do.
8. Arched ribs with braced spandrils.
9. Suspension with lattice stiffening girders.
10. Suspended girders.
11. Straight link suspension.

The comparative weights of the above constructions, both in iron and steel, will be investigated; but we shall first complete the necessary calculations for obtaining the weight of each of them in iron before we introduce the more novel material steel.

Commencing with the most unfavorable type for long-span railway bridges which it will be necessary for us to investigate—the box girder with plate webs—we might, without any preliminary calculation, and with a very little amount of consideration, foretell the uneconomical results which must necessarily follow the distribution of metal in such an unsuitable form. Thus, the economical depth will be much less, and consequently the sectional area of flange required for a given load will be much greater, than in any other type of girder. Again the amount of metal required to prevent the buckling of the deep thin plates would be nearly sufficient to form the struts of a lattice girder; therefore, the effective duty of such a web will be little more than the resistance it offers to tensional strains. But these strains may be more economically disposed of by means of lattice bars than by a solid plate; for, in the first case, the section may always be made proportional to the strain on the individual bar, whereas in the latter instance a certain minimum thickness of plate must be carried throughout, thus involving a waste of metal throughout nearly the entire length of the girder. Now the mass of a plate girder for each square inch sectional area of flange at centre we have found to be equal to $2(Sx+yd)$. Taking the weight of a bar of iron 1 ft. long and 1 in. square at .03 cwt., the above mass multiplied by .03 will give the weight of the girder in cwts. for each square inch section of flange. But the weight multiplied by ⅛ span will give the moment μ:

$$\mu = .03 \frac{S^2 x \times Syd}{4}.$$

Again, the strain, t, in cwts. per square inch resulting from the weight of the girder itself will be $\frac{\mu}{d}$; hence, since $Sx = Syd$ when d is the most economical depth, we have:

$$t = \frac{.015\ S^2 x}{d}.$$

Now, the influence of the weight of the web is the most important element in determining the proper depth for a girder, because whilst all the disturbing influences affecting the flanges also affect the web, there is, in addition, another element introduced, namely, the limiting thickness below which the plates may not be reduced; this is never taken at less than ¼ in., and in situations not easily accessible for the purpose of painting this thickness should be increased to ⅜ in.; again, when the load per foot is large, a thicker plate is usually employed. Now, the effective horizontal section of a web of uniform thickness, taking, as in the case of the flanges, a reduced strain to compensate for the loss of section through the rivet-holes, will, since the strain increases uniformly from the centre to the ends, and the span, S, being in feet, be equal to $12 \text{ S} \times \frac{1}{2} \times \frac{1}{2}$ thickness $= 3 \text{ S} \times$ thickness; consequently, as the least thickness of plate is ¼ in., the least effective horizontal section of any web will be ¾ S; and there can be but a small reduction in the weight of a web of uniform thickness, whatever the lightness of the load.

As far as the web is concerned, there would obviously be a practical advantage in making the depth of a girder small in proportion to the span and load. Thus, in shallow girders heavily loaded the gross average thickness varies from twice the net for short to two and a half times for long spans; but as the small depth is a disadvantage to the flanges, the determination of the depth at which the joint weights of the flanges and web would be a minimum is the problem to be solved. Now, W being the distributed load, the other notation remaining as before, the sectional area (a) at the centre of the flange, in square inches, will be $a = \dfrac{\text{WS}}{8d'\text{T}}$, which amount will also represent the actual horizontal section required in the half web; but we have seen that in practice this latter area is never less than ⅔ S; consequently, the value of a for the web must never be taken at *less* than that amount.

As we know the mass, and consequently the weight, of a girder to be proportioned to $a\,(Sx+y_1d)$, it is only necessary now to ascertain to what extent the limiting value of a for the web will affect the question.

Now, taking a web of uniform thickness, and adopting the highest value of y_1 and the lowest of x, it is obvious that if the mass of the flanges exceed that of the web, the depth must necessarily be too small, since an increased depth would similarly affect the weight of the web directly, and the flange inversely; thus, assuming the mass of the flange to be that of the web as 6 : 4, the sum being 10, then, if the depth were increased ¼, the mass of the web would be $\frac{4\times 5}{4}=5$; and that of the flange $\frac{6\times 4}{5}=4.8$, the sum being 9.8. Again, taking the highest value of x and the lowest of y_1, we can arrive in the same manner at the maximum depth. If the web be not of uniform thickness, it is even more apparent that an excess in the mass of the flanges over that of the web indicates deficient depth, since increased depth would involve a proportionally less increase in the weight of the web. We have, therefore, $axS = a_1y_1d$, and we know that a_1 can never be less than ⅔ S, and never be more than $\frac{WS}{8dT}$. Now, supposing that the thickness of the web might be reduced indefinitely, then, in order for the mass of the flange to be equal to that of the web, the span must obviously be to the depth as $y_1 : x$; thus, theoretically, the flange and web will be of similar weights when the depth equals ⅔ of the span. Although we cannot reduce the thickness of the web below a certain amount, the variations in the value of y_1 are too small to affect materially the economy of using as thin a web as possible. Taking a web ¼ in. thick, therefore, we have $axS = \tfrac{4}{3}Sy_1 d$, and $a = \dfrac{WS}{8dT}$; therefore, $WSx = 6d^2 Ty_1$, and $d^2 = \dfrac{WSx}{6Ty_1}$.

We have previously observed that a thicker plate than ¼ in. is desirable if the load per foot exceeds a certain amount, the limits of which must of necessity be determined somewhat arbitrarily; this will, however, only affect the value of the constant (6).

Now, taking a useful load of 35 cwts. per foot to be carried by each girder, which will be a sufficient approximation to the truth for our present purpose, we have:

$$W = \frac{35\,TS}{T-t}.$$

Again, taking ¼ in. bare as the minimum thickness of plate allowable in the construction of a box girder 300 ft. span, and ½ in. full as that necessary for a similar girder 600 ft. span, we obtain for the 300 ft. span the value of the constant $(c) = 10$; and for the 600 ft. span, $c = 26$. Generally, we may put $c = 3.5 + \dfrac{W}{10\,S}$. From existing structures we deduce the values of x and y to be respectively $x = .93$, and $y = 5.4$ for short spans, increasing in a certain ratio with the span, on account of the extra amount of bracing required; say, $y' = 5.4 + .002\ S$.

We have, therefore, the strain in cwts. per square on the flange of a box girder due to its own weight:

$$t = \frac{.015\ S^2 x}{d},$$

when $x = .93$, and $y' = 5.4 + .002\ S$.

$$d = \sqrt{\frac{W S x}{T\ c y'}}\ ;\ \ W = \frac{35\ T S}{T - t},\ \text{and}\ c = 3.5 + \frac{W}{10\ S}.$$

Substituting and reducing, we obtain, when $T = 80$ cwts. per square inch:

$$t = \sqrt{162 n + \frac{n^2}{4}} - \frac{n}{2}.$$

when $n = \dfrac{2640\ S^2 + S^3}{23,700,000}.$

Applying this formula to the given spans, we obtain the results shown in the following Table:—

Span in feet.	Strain in cwts. per sq. in.	Economic depth.
300	37	⅑ span
400	48	
500	58	
600	68	
700	77	
800	85	⅛ span

As we have now before us the strains per square inch on the girders resulting from their own weights, we can, by the methods already shown, at once obtain the weight of iron required to carry a given load as it will be expressed in terms of that load by $\frac{t}{T-t}$, the values of which for the different spans are shown below:

Span in feet.				Multiple.
300	×	$\frac{17}{43}$	=	.86
400	×	$\frac{16}{32}$	=	1.5
500	×	$\frac{58}{22}$	=	2.6
600	×	$\frac{68}{12}$	=	5.6
700	×	$\frac{77}{3}$	=	25.6
800				∞

We shall defer for the present any consideration of the probable load to which the above spans are liable in railway bridges, and necessarily also, of the actual weight of iron required in each instance. It will be found more convenient to treat our type girders collectively, with reference to the load; we shall, therefore, first advance them all as far as the preceding stage.

In the course of our present investigations, we shall constantly find instances where theoretical advantage in form is overruled, and more than neutralized, by some practical disadvantage incidental to the construction of the girder. The type we have already considered is a case in point. Theoretically the plate web girder requires less metal than any other form, and next to it in economy ranks the lattice girder with bars at the angle of 45°. Now, in practice we find these conditions to be precisely reversed, the plate web ranking lowest, whilst, as we shall hereafter show, the lattice girder is superior to that type alone in the scale of economy.

The practical advantage of the lattice girder over the plate is due to the greater depth which may economically be employed in the former system, and not to the smaller quantity of metal required in girders of equal depth in both instances. The want of a correct appreciation of this fact, or the way in which it is commonly ignored, is evidenced but too forcibly in the massive stunted lattice girders so prevalent on English railways; and it is no just cause for surprise that girders of such proportions should not compete successfully on the score of economy with the spider-like trussing of American lattice bridges.

Although the stiffness of a lattice girder is less than that of a box girder of similar depth, the stiffness of the type lattice girder will be greater than that of the type box girder, on account of the greater depth obtained in the former construction.

Adopting a similar mode of procedure to that already exhibited in the instance of the box girder, we have for

Type 2.—*Lattice Girder.*

Mass of girder for each square inch section of flange at centre proportional to $2(Sx + 2yd)$; which, multiplied by $\frac{1}{4}$ span and by .03 cwt., will give the moment:

$$\mu = .03 \frac{S^2 x + 2 Syd}{4}.$$

But $\frac{\mu}{d}$ = strain per square inch, and since $Sx = 2yd$, the economic depth will be $d = \dfrac{Sx}{2y}$. By substitution we have, therefore, the strain in cwts. per square inch, due to the weight of the girder itself:

$$t = .03\, Sy.$$

Availing ourselves, as before, of the data afforded by existing girders, we have $x = .93$, and $y = 2.7 + .001\, S$; hence

$$t = .081\, S + .00003\, S^2.$$

It appears also that the economic depth diminishes from $\frac{1}{6.5}$, the span at 300 ft., to $\frac{1}{7.5}$, the span at 800, thus showing the operation of a different law to that exhibited in the instance of the box girder.

The following Table shows the strain in cwts. per square inch, resulting from the weight of the girder itself in each instance:—

Span.	Strain in cwts. per sq. in.	Economic depth.
300	27	$\frac{1}{6.5}$
400	37.2	
500	48	$d = \dfrac{Sx}{2y}.$
600	59.4	
700	71.4	
800	84	$\frac{1}{7.5}$ span.

Since the weights of the girder itself and of its load are similarly distributed, the amount of the former weight will be the multiple $\dfrac{t}{T - t}$ of the latter. The results for the several spans are given below:

Span.			Multiple.
300	$\frac{27}{53}$	=	.51
400	$\frac{37.2}{42.8}$	=	.87
500	$\frac{48}{32}$	=	1.5
600	$\frac{59.4}{20.6}$	=	2.83
700	$\frac{71.4}{8.8}$	=	8.3
800		=	∞

Type 3.—*Bowstring Girders.*

The bowstring girder consists of an arched rib carrying the load, and a tie, instead of the usual abutments, resisting the thrust of the arch, and preventing the tendency to spread at the feet. If the line passing through the centre of gravity of all the cross sections of the arched rib corresponds in position with the curve of equilibrium due to the distribution of the load, the only connection necessary between the arch and the tie will be such vertical ties as may be required to transmit the weight of the tie and its insistent load to the arched rib. But, as the condition of stability of the arched rib requires that the curve of equilibrium of the load, as transmitted to it, should correspond in form with the arch itself, it follows that, if the distribution of the load be different to that required, the mere insertion of vertical ties capable of transmitting the load in direct lines only will not be sufficient. It will obviously be necessary to supply diagonal members capable of transmitting the load to the points in the arch where it is required to effect compliance with the conditions of equilibrium.

It appears, therefore, that, as the ordinary segmental

arched rib of a bowstring bridge corresponds in form very nearly with the parabolic curve of equilibrium due to a distributed load, no diagonal ties are required; but it does not necessarily follow that if such ties be inserted the strain on them would be *nil;* in point of fact, the deflection of the girder would communicate a certain amount of strain to those members. If, however, the load be rolling, it will be absolutely necessary to supply such an amount of bracing as may be required to effect the equal distribution of the load on the arched rib, whatever may be its actual position on the platform of the bridge.

In either case, however, the quantity of metal required will be small compared with the amount absorbed in the corresponding "web" portion of a parallel girder. We may, therefore, make a bowstring girder of greater depth than we could economically use in the box or lattice construction. That is the reason why we can in every instance construct a bowstring girder of no greater weight than a plate or lattice girder of corresponding strength; and why, in cases where the dead load is great, or, in other words, when the span is long, we may even construct it with a much smaller amount of metal, notwithstanding the heavy theoretical disadvantage the bowstring girder labors under in the diminished depth towards the ends, and the consequent increase, instead of diminution, of sectional area of the arched rib at those points.

The stiffness of a bowstring girder will be less than that of a lattice girder of similar depth and strength. If there were no diagonal members in the construction, the deflection of the arched rib alone, supposing the abutments fixed, would be nearly equal to that of the lattice girder; whilst, as the abutments are movable to the extent of the extension of the tie, that amount of deflection would be about doubled. The introduction of bracing would diminish this deflection in proportion to the strength and adjustment of its several parts. With the type bowstring girder this condition and the increased depth will diminish the deflection to about the same amount as that obtained in the instance of the type lattice girder.

From what we have already observed, it may easily be deduced that the weight of iron required in a bowstring bridge to sustain a given load will be governed chiefly by the amount, but partly also by the nature, of the load. In a railway bridge the load is of a mixed character, consisting of dead and rolling elements in varying proportions. We shall therefore first deal with the two extreme conditions—all dead and all rolling loads—and ascertain the comparative quantity of metal required in each. To obtain that necessary to carry a mixed load, we shall merely combine these amounts in the same proportions as the dead and rolling elements obtain in the total load. This, of course, is not a correct method theoretically, as we are adding together + and — strains on some of the members. We have, however, in this instance, as in several others, in accordance with our professed intention of avoiding all complications, preferred keeping the question in its simplest form; and as the required corrections are effected by means of the coefficient y, the final result obtained will be correct, although the process by which it was arrived at is not quite so.

Dead Load (*uniformly distributed*).

The mass of the tie will be as a S simply, and the mass of the arched rib will be that of the tie, together, with a certain additional amount due to the shearing strain transmitted through the arch. The shearing force in action consists of the entire weight of the load, less a certain proportion of it which may be resting immediately on the piers without being first transmitted through the arched rib. Now, instead of ascertaining the increased sectional area required towards the ends of the arched rib in consequence of the combined action of the uniform horizontal force and of the uniformly increasing shearing force, and multiplying the mean sectional area thus determined by the length of the curved rib for its mass, we may approximate to the same result very nearly, and by a very simple process, if we keep the two forces in action distinct throughout, and deal with their masses separately. Thus we shall imagine the arched rib to be replaced by a horizontal member to resist the horizontal force, and by a vertical member equal in height to the depth of the girder to resist the shearing force.

The mass of the horizontal member will, of course, be the same as that of the tie $= a$ S, and the mass of the vertical member will, if we take $\frac{7}{8}$ of the gross load as the amount transmitted through the arch, be to that amount as $\frac{7d^2}{S}$: S. The total mass of the arched rib and tie will therefore be as

$$a \left(2S + \frac{7d^2}{S}\right).$$

Again, the mass of the vertical members transmitting the weight of the tie and its insistent load to the arched rib may, for our purpose, be taken as equal to $\frac{2d^2}{S}$; consequently, if x and y be coefficients for the horizontal and vertical members respectively representing the ratio of the metal used in practice to that theoretically required, the total mass of the bowstring girder for a uniform load will be

$$a \left(2Sx + \frac{9d'y}{S}\right).$$

Again, the mass multiplied by $\frac{1}{8}$ span and by .03 cwt., will give the amount in cwts. for each square inch sectional area of tie; hence for 1 square inch,

$$\mu = .03 \left(\frac{S^2x}{4} + \frac{9d^2y}{8}\right).$$

But $\frac{\mu}{d}$ = strain in cwts. per square inch; and since economic $d^2 = \frac{2S^2x}{9y}$, we have $d = \frac{7}{15} S \sqrt{\frac{x}{y}}$; and the strain in cwts. per square inch due to the weight of the girder itself equal to

$$t = .03 \frac{15}{14} Sx \sqrt{\frac{y}{x}}.$$

Rolling Load.

As the maximum strains on both the arched rib and the tie occur when the rolling load entirely covers the bridge, it is obvious that the mass of those members will be the same as for a distributed load

$$= a\left(2S + \frac{7d^2}{S}\right).$$

The *web* portion, however, as we have already shown, will have a different duty to perform; and it is necessary now to ascertain to what extent this will affect the mass of the girder.

Now, it may be shown that the maximum horizontal strain to which the diagonals of a bowstring bridge are liable is the same for each bar; and that, assuming 10 bays of diagonals, the strain in each instance will be about $\frac{1}{10}$th of the maximum horizontal strain on the arched rib or tie, due to the rolling load uniformly distributed. The mass, therefore, of the horizontal components of one complete set of 10 diagonals will be $a\frac{S}{10}$.

Again, the mean square of the verticals being $\frac{d^2}{2}$ nearly, and having assumed 10 bays, the mass of the vertical components of the same set of diagonals will be $a\frac{5d^2}{S}$ nearly; consequently, the total mass of one set of diagonals will be:

$$a\left(\frac{5d^2}{S} + \frac{S}{10}\right) \text{nearly.}$$

As we have assumed two sets of diagonals, one crossing the other, and one set of struts, the total mass of the *web* portion of the girder will be

$$a\left(\frac{14d^2}{S} + \frac{S}{5}\right).$$

Taking the coefficients x and y of the same value as before, we have the mass of the bowstring girder for an entirely rolling load equal to

$$a\left(\frac{11Sx}{5} + \frac{21d^2y}{S}\right).$$

Now, the moment in cwts. equals the mass multiplied by ⅛ span and by .03 cwts.; hence we have for 1 square inch area of tie:

$$\mu = .03\left(\frac{11S^2x}{40} + \frac{21d^2y}{S}\right).$$

But $\frac{\mu}{d}$ = strain in cwts. per square inch, and $d^2 = \frac{11S^2x}{105y}$;

hence $d = \frac{10}{31}S\sqrt{\frac{x}{y}}$; and, by substitution, the strain due to the weight of the girder equals

$$t = .03\frac{17}{10}Sx\sqrt{\frac{y}{x}}.$$

Mixed Load.

It appears, therefore, that the strain per square inch, due to its own weight, on a bowstring girder constructed to carry a dead load only, will be to that occurring on an equally strong girder constructed for a rolling load of similar amount as $1\frac{5}{6} : 1\frac{7}{4}$. Now, excluding the weight of the girders, the load to be carried by a railway bridge may be considered as consisting of ⅔ths rolling and ⅓th dead load; consequently, the equivalent fraction for the mixed load will be $1\frac{7}{6} \times \frac{1}{3} \times 1\frac{9}{4} \times \frac{1}{3} \times \frac{108}{70}$; or, taking the fraction $1\frac{5}{7}$ for the dead load as the unit of measurement, that for the rolling load will be 1.44 times the amount.

Putting $.03\frac{15}{14}Sx\sqrt{\frac{y}{x}}=a$, we have when $x=1.25$, and $y=3.2+.002S$:

$$a=.04S\sqrt{2.56+.0016S};$$

and the strain in cwts. per square inch due to the weight of a bowstring girder for a railway bridge will be

$$t=\frac{1.44aT}{T+.44a}.$$

Taking the limiting strain, $T=80$ cwts. per square inch, and arranging the results as before in a tabular form, we have:

Span in feet.		Strains in cwts. per square inch.	Depth.
300	=	27	⅛ span
400	=	35.5	
500	=	43.5	
600	=	52	
700	=	60	
800	=	67	
900	=	75	
1000	=	80	⅛ span

The multiple $\frac{t}{T-t}$ will have the following values:

Span in feet.			Multiple.
300	$\frac{27}{53}$	=	.51
400	$\frac{35.5}{44.5}$	=	.79
500	$\frac{43.5}{36.5}$	=	1.19
600	$\frac{52}{28}$	=	1.86
700	$\frac{60}{20}$	=	3.
800	$\frac{67}{13}$	=	5.15
900	$\frac{75}{5}$	=	15.
1000		=	∞

Type 4.—*Straight-link Girders.*

We are not aware that any example of the above mode of construction exists in this country. The nearest allied to it, perhaps, is Brunel's Chepstow bridge, and even that structure may be more properly referred to our second type, as in reality it is little more in principle than a three-bay lattice bridge, the most noticeable feature, and the one redeeming the design from the charge of extravagance, being its great depth, amounting to about one-fourth the span.

In America, on the other hand, the straight-link girder, under the name of the Bollman truss, appears to meet with general approval; and, as there are several large bridges of the class erected in that country, we may consider the principle to have been fairly tested as to its practical capabilities.

Theoretically, our present type is the heaviest form of girder we have yet considered, and in the discussion at the Institution of Civil Engineers following Mr. Zerah Colburn's paper on American iron bridges, it was on that evidence condemned as uneconomical. We have, however, advanced sufficiently with our investigations to be satisfied how fallacious conclusions must be which rest on so uncertain a basis as mere theoretical considerations. All our types as yet have occupied precisely reverse positions in the scale of economy to that indicated for them respectively by theory; we may not, therefore, be surprised if we find this—the lowest in the scale—positively heading its competitors.

In principle the straight-link girder is more nearly

allied to the bowstring than to any other system. In both instances we have one straight member of uniform section throughout—the tie and the boom—opposed by a member of greater length and of increased section towards the piers—the arched rib and the collection of ties. We found the weight of the bowstring girder to vary considerably with the character of its load, and it will be seen hereafter that the same element influences the weight of the straight-link girder, the conditions, however, being reversed. From the nature of the trussing in the present case, a rolling load will be more economically dealt with than will the load due to the weight of the girder itself, whilst it will be remembered the former load operated disadvantageously on the bowstring girder. It follows from this that if the short-span bowstring girder has any advantage over our present type, it will maintain a still greater advantage in the long-spans, whilst on the other hand, if the straight-link girder excels the former systems for the short spans, it by no means follows that it will be able to compete with it for the long spans. Indeed, theoretical deductions show it to be otherwise; and as in this instance they are not overruled by practical considerations, the fact of the joint moment of all the bars being $1\frac{1}{4}$ times that of the arched rib of equal strength must detract from the economy of the straight-link girder, as compared with the former system for uniform loads.

With reference to the comparative stiffness of this description of bracing, it was remarked in the discussion at the Institution of Civil Engineers previously alluded to, that diagonals of great inclination were free to deflect on a curve struck from one of the ends as a centre, and the other end as a radius; and that, as the curve so described would coincide in practice for a considerable distance with a straight line, there would be little or no resistance to deflection. Now this conclusion is palpably false, as the deflection would be inversely as the angle included between the given pair of ties, and not depend upon the inclination of either of them to the horizon. Thus, if the two bars be at right angles to each other, it is immaterial as regards deflection whether both of the bars be at the angle of 45° to the horizon, or whether one be vertical and the other at the *flattest* possible angle, in fact horizontal, since with equal depths and unit strains the deflections would be similar.

As the most obtuse angle included between any two bars occurs at the centre of the span, the deflection will be greatest at that point, and will be less towards the ends, as in other structures; the amount, however, in this instance will be about double that occuring on a lattice girder of similar depth and strain per square inch. In this respect it resembles the bowstring girder, but the practical disturbing elements reducing this double amount in the case of the bowstring girder will not be obtained in the present instance.

There is a peculiarity, however, connected with the deflection of this girder, which should not be passed over without notice, as it may be of grave practical importance. As each pair of ties acts independently, affecting the others indirectly only through the medium of the boom, it follows that if any pair of ties have their full proportion of load, they will also incur a large moiety of their full deflection, which deflection will not be shared in any perceptible degree by the remaining unloaded portions of the bridge. The practical effect of this condition is, that as the rolling load advances along the bridge, it will, so to speak, break the back of the boom at the vertical, springing from the nearest adjacent unloaded pair of ties.

With the 300 ft. span bridge, for example, the load being half over, the deflection at the centre would be about 3 in., whilst, if the bridge consisted of ten bays, at a point 30 ft. off, the deflection would only be 1 in. In the length of 30 ft., therefore, we have to dispose of a difference in level of 2 in., about; and this will necessitate either an elastic boom subject to transverse strain, or else one jointed at each vertical. If we attempt to get over the difficulty by the insertion of bracing to equalize the deflection, as is sometimes done in the American bridges, there will be little hope of attaining an economical structure.

Having pointed out the foregoing practical difficulty in the construction of the straight-link girder, we shall assume it to be surmounted, without producing any abnormal strain on the boom, and without any extra provision of bracing, and shall now proceed to ascertain the quantity of metal required on that hypothesis.

Dead Load.

The mass of the boom will be proportional to $a\,S$, and the mass of the ties will be that of the boom, together with the additional amount due to the transmission of the shearing strains through those members.

We shall deal with the masses required to resist the horizontal and vertical strains separately, as we did in the instance of the bowstring girder.

Now, taking the cluster of ties collected at each pier it may be shown that the horizontal strain on each will be proportional to the ordinates of a parabola, of which the highest represents the horizontal strain on the centre pair of ties. The mean height of all the ordinates of a parabola being equal to $\frac{2}{3}$ of the height, it follows that the mean horizontal strain on each bar will be $\frac{2}{3}$ of the amount on the centre bars. But the moment of the load on the centre pair of ties is equal to $W\frac{S}{4}$; consequently the mean moment of all the ties will be $\frac{2}{3}$ of that amount $= W\frac{S}{6}$. It appears therefore, with similar depths, loads, and unit strains, the sectional area of the *flange* portions of a straight-link girder will be $\frac{2}{3}$ of that necessary in either of the types we have yet considered.

The mass of the horizontal components of the ties being $a\,S$, the mass of the vertical components will, if we take $\frac{7}{8}$ of the total load as transmitted through them, be to that amount as $\frac{21d^2}{4S}$: S. The total mass of ties and boom will therefore be $= a\left(2\,S + \frac{21d^2}{4S}\right)$.

But we must also provide vertical members to support the boom and ties at certain intervals, the mass of which members may be taken as equal to $\frac{6d^2}{S}$; therefore, taking x and y coefficients for the horizontal and vertical members respectively, the total mass of metal required in the construction of a straight-link girder for a dead load will be

$$a\left(2Sx + \frac{45\ d^2y}{4S}\right).$$

Taking $a = 1$ square inch, the mass multiplied by .03 cwts. and by $\frac{1}{8}$ span will equal the moment in cwts.

$$\mu = .03\left(\frac{S^2x}{3} + \frac{15d^2y}{8}\right).$$

But $\frac{\mu}{d}$ = strain in cwts. per square inch; and since economic $d^2 = \frac{8\ S^2x}{45\ y}$, we have $d = \frac{S}{2.37}\sqrt{\frac{x}{y}}$; and the strain in cwts. per square inch, resulting from the weight of the girder itself, equal to

$$t = .03\left(1.58\ Sx\sqrt{\frac{y}{x}}\right).$$

Rolling Load.

As the maximum strain on the ties and boom is attained when the bridge is entirely loaded, the mass of those members will be the same as before,

$$= a\ 2S\left(+\frac{21d^2}{4S}\right);$$

and as we have assumed the platform of the bridge to be at the level of the bottom of the girder, no additional metal will be required to complete the girder for the rolling load. Therefore, x and y being the coefficients as before, the total mass will be:

$$a\left(2\ Sx + \frac{21d'y}{4S}\right).$$

And, for 1 square inch area of boom the moment, in cwts. = mass × .03 cwts. × span will be:

$$\mu = .03\left(\frac{S^2 x}{3} + \frac{7d^2 y}{8}\right).$$

But $\frac{\mu}{d}$ = strain in cwts. per square inch; and since economic $d^2 = \frac{8 S^2 x}{21 y}$, $d = \frac{S}{1.62}\sqrt{\frac{x}{y}}$; and the strain in cwts. per square inch due to the weight of the girder itself will be:

$$t = .03\left(1.08 S x \sqrt{\frac{y}{x}}\right).$$

Mixed Load.

Putting $.03\left(1.58 S x \sqrt{\frac{y}{x}}\right) = a$, we have when $x = 1.25$, and $y = 3.2 + .002\ S$:

$$a = .059\ S\sqrt{2.56 + .0016\ S};$$

and the strain in cwts. per square inch due to the weight of a straight-link girder for a railway bridge will be:

$$t = \frac{.7 a T}{T - 3a}.$$

Taking the limiting strain, $T = 80$ cwts. per square inch and substituting the spans, we obtain the following results:

Span in feet.	Strain in cwts. per square inch.	Depth.
300	23.1	¼
400	33.5	
500	45.0	
600	57.7	
700	72.7	
800	92.5	⅛

The multiple $\dfrac{t}{T-t}$ will have the following values:

Span in feet.			Multiple.
300	$\dfrac{23.1}{56.9}$	=	.4
400	$\dfrac{33.5}{46.5}$	=	.72
500	$\dfrac{45}{35}$	=	1.28
600	$\dfrac{57.7}{22.3}$	=	2.61
700	$\dfrac{72.7}{7.3}$	=	9.97
800		=	∞

We have now arrived at the conclusion of what may be considered the first stage of our investigations. Our type girders, Nos. 1, 2, 3, and 4, are all independent structures, carrying their loads without any extraneous assistance, the only requisite being a supporting pier at each end capable of bearing one-half the maximum load on the bridge. As, in our opinion, all justifiable modes of constructing independent girders may be referred to one or the other of the preceding types, and will be included within those limits, we shall now proceed to the second stage of our inquiry, which refers to structures whose stability depends upon some support beyond that afforded by the simple pier. We shall first consider those structures which—although dependent upon external assistance—produce only a vertical pressure on the piers; and secondly, the systems whose stability is governed by the power of the piers, or abutments, to resist horizontal, as well as vertical, forces.

The first division will comprise Types 5, 6, and 7, that is, two kinds of cantilevers and the continuous girder; and the second division will include the remaining type structurss.

Type 5.—*Cantilever Girders of uniform depth.*

In appearance the type we have now to consider is identical with the independent lattice girder, and this identity is not merely apparent, but does, in fact, obtain to a great extent in the web; but a little consideration will show that the flanges are placed under entirely different conditions.

In the independent girder of uniform depth, and with a distributed load, the strain on the flanges will be greatest at the centre of the span, and less elsewhere, in proportion to the ordinates of a parabola; whilst in a similar cantilever girder, although the maximum strain will be the same in *amount* as before, it will take place at the piers, and will diminish towards the centre in proportion to the co-ordinates of the same parabola; at the middle of the span the strain, therefore, will vanish. It follows that, in the independent girder, the theoretical mass of flange required will be, to the mass obtained by multiplying the sectional area into the length, as the area of a parabola: the area of the inclosing rectangle, that is, as $\frac{2}{3}$: 1; whilst in the cantilever the proportion will be as the area of the complement of a parabola to the same rectangle, that is, as $\frac{1}{3}$: 1. The mass of metal required theoretically to form the flanges of a cantilever will therefore be only one-half of that necessary in a similar independent girder, the load in both instances being uniformly distributed.

This proportion, however, does not represent the whole of the advantage accruing to the former system, as the *moment* of the flanges will be diminished in a much higher ratio than the *weight*.

The centre of gravity of the semi-parabola, representing the mass of metal required in the flanges each side of the centre of an independent girdle, being $\frac{5}{16}$ of the span from the pier, the moment of the flange will be proportional to $\frac{2}{3} \times \frac{5}{16} = \frac{5}{24}$; whilst the centre of gravity of the complement of the semi-parabola being $\frac{1}{8}$ of the span from the pier, the moment of the flanges of a cantilever girder will be proportional to $\frac{1}{3} \times \frac{1}{8} = \frac{1}{24}$. The ratio of 5 to 1 indicated by the preceding theoretical considerations is so high that we may be sure, after allowing an ample margin for all possible contingencies, a balance will still remain in favor of our present type, which must tell with considerable effect in the economy of long-span bridges, where a large proportion of the load consists of the weight of the girdle itself.

Of course, whether the extra metal required in the adjacent spans, according to this system, may equal or exceed the saving effected in the main span is another question, which does not concern us at present.

The maximum deflection of the cantilever girder

will obviously be partly governed by the adjacent spans. If the side spans be one-half the main span, the deflection at the centre of the latter will be from $1\tfrac{1}{4}$ times to double that of the independent girder, of equal depth and unit strains. The unequal deflection of the two halves of the main span, due to the passage of a rolling load, presents no practical difficulty, as a very simple connection may be contrived, admitting free vertical deflection and longitudinal expansion, but at the same time resisting any tendency to lateral movement.

Now the mass of the girder for each square inch area of the maximum cross section of flange will be the same as for Type 2, that is:

$$2(Sx + yd).$$

Since the mass multiplied by .03 cwts. and by $\dfrac{S}{b}$ = moment in cwts., we have:

$$\mu = .03\frac{2S^2 + 4Syd}{b}.$$

But $\dfrac{\mu}{d}$ = strain in cwts. per square inch (t); and since economic depth = $\dfrac{Sx}{2y}$, we have:

$$t = .03\frac{8Sy}{b}.$$

The value of b will vary according to the distribution of the load; for an uniform load $b = 8$, and we must now ascertain its values for the various distributions of load obtained in railway bridges of various spans.

Now, if the load to be carried be uniformly distributed, the value of b for the portion of the load consisting of the weight of the girder itself will be as follows:

Web $\quad=12$ } mean theoretical value $b=14$.
Flanges$=16$ } " practical " $b=11$.

Again, if the load consists of the weight of the girder alone we have:

Web $\quad=16$ } mean theoretical value $b=18$.
Flanges$=20$ } " practical " $b=15$.

It follows, therefore—T being the limiting strain in cwts. per square inch—the value of b, corresponding to the required strain, t will be:

$$b = \frac{15t + 11(T-t)}{T}.$$

Substituting this value in the former equation, and taking $T=80$ cwts., $y=2.5+.001S$, and $x=.6$, we obtain:

$$t = \sqrt{12S + .005 S^2 + 12100} - 110;$$

which equation gives the following results:

Span in feet.	Strain in cwt. per sq. in.	Depth.
300	17.2	$\frac{1}{4}$
400	23.2	
500	29.2	
600	33.1	
700	41.7	
800	48.0	
900	54.4	
1000	60.5	
1100	67.0	
1200	73.8	
1300	80.0	$\frac{1}{12}$

We have already observed that for a distributed load, such as the weight on the platform of a railway bridge, $b = 8$; and we have also found its value for the girder weight required to carry this load at the various spans. We can, therefore, at once obtain the weight of iron required in the construction of the main girders, as it will be the multiple of the load expressed by the equation:

$$\text{Multiple} = \frac{tb}{8(T-t)}.$$

Span in feet.	Multiple.
300	.39
400	.61
500	.88
600	1.23
700	1.64
800	2.5
900	3.6
1000	5.43
1100	9.25
1200	21.6
1300	∞

Type 6.—*Cantilever Lattice Girder, varying economic depth.*

The most superficial examination of the method and results of our investigations concerning the cantilever lattice girder of uniform depth could hardly fail to suggest a desirable modification in its outline. Thus, if we lay off 300 ft. span and plot the economic depth at each end, and then, adding 50 ft. to each end, plot the economic depth for 400 ft. span, we shall, if we carry on the process up to the limiting span, and connect the various plotted heights by lines, obtain a curve the ordinates of which will represent the economic depth for the different sections of the girder, and consequently, if there be no new condition introduced, of the entire girder.

But the alteration of the top flange from a straight line parallel to the bottom flange, to a curved line inclined to the latter, *does* introduce a new element into the case, as a portion of the shearing strain will now pass through the top flange, and to that extent, of course, the *web* will be relieved of its strain.

Now an examination of our last investigation will show that the depth of the girder, and, necessarily, to a great extent the weight also, is governed by the mass of the *web* portion. It follows, therefore, that the more we can reduce the strain on and, consequently, the mass of, those portions, the greater will be the economic depth, and, within certain limits, the smaller the total mass of metal required in the construction of the girder. It is not difficult, then, to see that we should employ every available economical means of reducing the strain on the *web* portions. We can, fortunately, effect this by the very simple process of giving the upper flange a downward curvature. The tendency of the tension member to pull straight will react on the long struts, and by the production of an initial tension reduce the mass of metal required for those members of the web; whilst, if we make the curved outline *include* the various economic depths, we shall arrive at a stiffer form of girder than before.

It would be foreign to our present purpose, and

inconsistent with our avowed intention of viewing our subject in the broadest possible light, were we to endeavor to deduce the precise amount of curvature which would give the most favorable general results. It will be a sufficiently near approximation to the correct average proportions, if we assume the depth at the centre of the girder to be one-fourth that at the ends, and the curvature of the top flange to be the segment of a circle passing through those three points.

The form of bridge to which we have been thus, as it were, irresistibly driven is, we believe, almost identical in general and outline proportions with the structure designed by Mr. Fowler to span the 600 ft. centre opening, and the two 300 ft. side openings, of his viaduct for carrying the proposed "South Wales and Great Western direct" railway across the Severn estuary.

It will be unnecessary to give a detailed investigation of this modified form of the cantilever lattice girder. The horizontal components of the diagonals will be, proportionally, the same as before; the vertical components, however, will be less, as a proportion of the shearing strain is transmitted through the curved top flange. The struts, again, will be lighter, on account of their diminished length, and, for the same reason, the flanges will be heavier. The centre of gravity of the mass of the entire girder being about the same as in the previous instance, we have merely to substitute new values of x and y in the equation already deduced for the parallel cantilever girder. Taking $x = .7$, and $y = 2 + .001S$, the strain in cwts.

per square inch due to the weight of the girder itself will be:

$$t = \sqrt{9.6S + .004S^2 + 12100} - 110,$$

which equation gives the following results:

Span in feet.	Strain in cwts. per sq. in.	Depth.
300	14	¼
400	19	
500	24	
600	29	
700	34.5	
800	40	
900	45.5	
1000	51	
1100	56.5	
1200	62	
1300	68	
1400	74	
1500	80	⅛

The multiple, as in last case, will be given by the equation:

$$\text{Multiple} = \frac{tb}{8(T-t)}.$$

Span in feet.	Multiple.
300	.3
400	.46
500	.65
600	.88
700	1.2
800	1.61
900	2.17

Span in feet.	Multiple.
1000	2.96
1100	4.15
1200	6.1
1300	11.2
1400	22.6
1500	∞

Type 7.—*Continuous Girder, varying economic depth (including Sedley's patent).*

Our last investigation shows us that with an unlimited supply of metal we may construct a cantilever bridge up to 1500 ft. span; it follows, as a corollary, that it is also possible to construct a cantilever, or bracket, 750 ft. projection, capable of carrying any required load at its extremity. For, suppose we support this load by a simple triangular frame, consisting of an inclined tie and a horizontal strut, then the weight of this frame will not produce a strain on either of those members, since the whole affair will be borne by the original cantilever as a portion of its uniform load; and consequently, it will be possible to construct this triangular bracket, or what amounts to the same thing, the original cantilever, of sufficient strength to carry any required load at its extremity. This being so, it necessarily follows that it will be practicable to support the two ends of either of our independent girders on the extremities of a pair of these cantilevers, as securely as if they rested on their original piers.

It is evident at once that if we *do* thus insert an independent girder between the two halves of a cantilever bridge, the limiting span of the entire structure will be equal to the sum of the limiting spans of the two systems of which the bridge is composed; that is, if we take our last type and the bowstring bridge, the limiting span will be 1500 + 1000 = 2500 ft. Now, what conclusion must we draw from this fact? It appears we may on this system, with a definite amount of metal, bridge an opening which could not be spanned by either of the other systems we have yet investigated with an *infinite* quantity; and the irresistible conclusion is that at the high spans a much smaller amount of iron will be required in the construction of a "continuous" bridge of this class than would be necessary in one constructed on either of the other systems. This theoretical deduction is fully corroborated by the indisputable economy obtained in the bridges on this principle erected under Sedley's patent. The only thing we have to determine, then, is the span at which this superiority will begin to manifest itself, and that, of course, will vary with the degree of economy obtained in practice in the other systems with which it is to be compared. In our present investigation—as we have in all instances supplied a very liberal amount of metal for the construction of the several types—it will be when the sum of the multiples for the cantilever bracket and bowstring, reduced to the equivalent value when measured over the whole span, is smaller in amount than the multiple of the cantilever bridge for the same span. The tables we have already calculated for the two systems of which this type consists will enable us to ascertain when this condition is obtained almost by inspection.

There is no practical difficulty in the construction of this compound structure calling for special notice. The bridge may be made in one connected length, and merely jointed at the points of contrary flexure occurring at the junctions of the bowstring with the cantilever, expansion being provided for at the piers in the usual way, or the bowstring may be slung from the ends of the cantilevers, and the expansion allowed to take effect at those points of the span. In the latter case, precautions must be taken to insure the maintenance of the strength of the horizontal bracing past those points, so as to check all tendency to lateral movement.

The deflection will, of course, be the sum of the deflections of the two smaller spans into which the bridge may be broken up. This will give a rather smaller proportional deflection than that appertaining to either system individually. Now let $a=$ the sum of the lengths of the two cantilevers, and let $b=$ the span of the centre portion, or the bowstring girder; then $a+b=$ span of the "continuous" girder. Also let n be the multiple corresponding to span a, and m the multiple for span b, given in the tables of multiples for types 6 and 3 respectively. Then the load per foot on the cantilever in terms of the *useful* load as before, reduced to the equivalent load per foot distributed over the entire span, will be $\dfrac{a(n+1)}{a+b}$.

In the same manner the load from the bowstring will

be $\dfrac{b(m+1)}{a+b}$; and assuming the weight on the end of the bracket as equivalent to double the amount distributed, the load from the "triangle" will be $\dfrac{2b(m+1)}{a+b}$. The sum of these amounts will be the mean equivalent load; and that amount, less the unit *useful* load, will be the mean equivalent multiple; therefore,

$$\text{Multiple} = \frac{a(n+1) + 3b(m+1)}{a+b}.$$

Substituting the valves of m and n obtained from Tables 3 and 6, we find, by the preceding equation, that in higher spans than 550 ft. the "continuous" girder is ligher than the cantilever. At 550 ft. span, then, the economic span of the centre portion on the bowstring girder $=0$; and we know that at 2500 ft. the economic span is also the limiting span $=1000$ ft.

Now $\dfrac{2500-550}{1000} = 1.95$; and it will be sufficiently accurate for our purpose if we assume the economic span of the centre portion generally to be:

$$\text{Economic span} = \frac{\text{Span} - 550}{2}.$$

Substituting this value of b in our former equation, we obtain the following results:

Span.	Multiple.
300	.3
400	.46
500	.65
600	.85
700	1.1

Span.	Multiple.
800	1.33
900	1.61
1000	1.84
1100	2.1
1200	2.49
1300	2.87
1400	3.37
1500	3.91
1600	4.58
1700	5.35
1800	6.47
1900	7.98
2000	10.42
2100	14.32
2200	21.14
2300	29.12
2400	50
2500	∞

Type 8.—*Arched Ribs with Braced Spandrils.*

With very few exceptions, all the earlier examples of what would, at the time of their erection, be considered long-span bridges are constructed on the principle of the arch; indeed, in comparison with that construction, all the types are in their infancy; and yet, perhaps, with reference to no other system does there exist so much indefiniteness and difference of opinion as to the real direction and amount of the strain at any given point in the structure. The explanation of this is simple enough, since, before we can attempt to determine the strains on an ordinary arch, we must make certain assumptions, the probable truth of which may be sufficiently proved in *our* opinion, but not so in the judgment of others. As we cannot absolutely demonstrate the truth of our hypothesis, any one else is, of course, at perfect liberty to make a different one, which may, very probably, give an entirely different result.

If, however, we have given the relative position of three points through which the centre of pressure on the arch passes, its position is defined at every other point, for just the same reasons that the radius of a curve is defined when it has to pass through three points. It follows that if we arrange the details of an arched rib so as to insure the centre of pressure passing through three known points, we shall be enabled to determine all the conditions with the same precision, and therefore to proportion the strength of the several members to the maximum strains occurring on each with the same ease as we do in the most elementary form of truss.

We have already adopted a similar method in the instance of the continuous girder, by first determining the most economical position for the points of contrary flexure, and then, by constructional arrangements, securing the constant position of those points under all conditions of loading. By this means we not only obtain a great theoretical advantage, but, by reducing the complicated and, in fact, almost indeterminate problem of the determination of the actual strains occurring on a continuous girder to the simple case of the independent girder, we were enabled in practice to effect an additional saving by proportioning the several members with much greater nicety to the maximum strains on each.

The desideratum in the design of an arched rib is, therefore, that the centre of pressure should, under all conditions of loading, pass through the same three given points. The most obvious way of effecting this is by making the arched rib movable on pivots at the centre and at each springing, thus hinging, as it were, the rib at those three points. If the frictional resistance to turning on these pivots be small—that is, if their diameters be small—the centre of pressure would always pass through their centres, which, for a symmetrical cross section of rib and uniform unit strain, should correspond with the axis of the rib at those points.

Again, in this design, expansion and contraction will merely produce a rise or fall of the arch at the crown, without any incidental strain; whereas if the rib were continuous there would have been additional strain with the consequent loss of metal on that score.

We shall therefore confine our investigations to the arched rib jointed at the centre and at each springing. This, of course, assumes intermediate piers of sufficient stability to take up the unbalanced thrust due to the rolling load, if the arch be one of a series in a viaduct. If the piers be too lofty to admit of this necessary provision, we should adopt an arched rib of an entirely different pattern, which will be referred to in considersng Type 10.

The problem reduced to its present dimensions is a very simple one, admitting of a definite solution; and on that account, partly, there is no reason why the æsthetically perfect form of the arch should not be economically employed in wrought-iron bridges.

The deflection of an arched rib will be very nearly the same as that of a lattice girder of similar depth and strain per square inch.

In considering the strains to which an arched rib is liable, it will be necessary to resolve the gross load into its two elements—dead and rolling; for in this instance, as in several previous ones, the strain will be to a great extent governed by the *nature* of the load as well as by its amount.

Dead Load.

With a dead load uniformly distributed, it is not necessary for the spandril fillings of the arch to possess any bracing power, for the same reasons that the diagonals of the bowstring girder were dispensed with under similar conditions. The mass of the arched rib will be the same as in the instance of the same girder:

$$= a\left(S + \frac{7d^2}{S}\right);$$

that of the verticals will be less, on account of their decreased length, the mass being about $\frac{ad^2}{S}$. Taking the coefficients, x and y, for the horizontal and vertical components of the strain, as before, the total mass of the arch for each square inch sectional area at crown will be:

$$Sx + \frac{8d^2y}{S}.$$

But the mass, multiplied by .03 cwts. and by ⅛th span, equals the moment:

$$\mu = .03\left(\frac{S^2 x}{8} + d'y\right).$$

Since economic $d'^2 = \frac{S^2 x}{8y}$, we have depth $= \frac{S}{2.83}\sqrt{\frac{x}{y}}$; and since the strain in cwts. per square inch (t) equals $\frac{\mu}{d'}$, we have for the dead load:

$$t = .03\left(.71\ Sx\sqrt{\frac{y}{x}}\right).$$

Rolling Load.

With a rolling load it will be necessary to introduce bracing between the arched rib and the "horizontal girder," of such strength that the moment of resistance to a traverse strain at any point of the spandrils may be equal to the bending moment at the same point, due to the unequal distribution of the load. We shall assume the depth of bracing at the centre of the span to be ⅛th the rise of the arch, which will give us an effective depth of ⅓d the rise at the point of maximum bending stress. This proportion will limit in all our examples the maximum strain on the arched rib to the same amount as it would be were the load entirely dead. The mass, therefore, will be the same as before:

$$= a\left(S + \frac{7d'^2}{S}\right).$$

Now the maximum bending moment will occur when the rolling load is half-way over the bridge, at which time it will, in terms of the bending moment on the entire span, be equal to $(\frac{1}{2})^3$. Since, however, the effective depth is only equal $\frac{1}{3}d$ the rise of the arch, the mass of metal required in the horizontal girder, or top member of the braced spandrils, will be equal to $(\frac{1}{2})^3 \div \frac{1}{3} = \frac{3}{8}$ aS.

The mass required for the sum of the horizontal components of the strain on all the diagonals will, if we provide a double set of 10 bays of bars, be equal to $\frac{9aS}{20}$. Again, the mean square of the verticals being $\frac{d^2}{2.6}$ about, it follows that the mass required in the vertical components of the same double set of diagonals will be $\frac{17d^2a}{S}$. The total mass of metal required in the arch for each square inch of sectional area at the crown will, therefore, taking coefficients x and y, as before, be equal to:

$$1.8Sx + \frac{24d^2y}{S}.$$

Since the mass multiplied by .03 cwts. and by $\frac{1}{8}$ span equals the moment, we have:

$$\mu = .03(.225S^2x + 3d^2y).$$

But $\frac{\mu}{d}$ = strain in cwts. per square inch (t); and economic $d^2 = \frac{1.8S^2x}{24y}$; hence $d = \frac{3S}{18}\sqrt{\frac{x}{y}}$, and by substitution

$$t = .03\left(1.62\, Sx\sqrt{\frac{y}{x}}\right).$$

Mixed Loads.

Assuming, as before, the useful load on a railway bridge to be composed of ¾ rolling and ¼ dead load, the mean coefficient will be $1.62 \times ¾ \times .71 \times ¼ = 1.4$; or, taking the coefficient for the dead load $= .71$ as the unit of measurement for the mixed load, it will be about double that amount. Hence the strain in cwts. per square inch (t) due to the weight of a girder for carrying a railway bridge will be:

$$t = .021\, Sx \sqrt{\frac{y}{x}} \left(2 - \frac{t}{T}\right).$$

Putting $a = .021\, Sx \sqrt{\frac{y}{x}}$, when $x = 1.25$, and $y = 3.5 + .002S$, we have:

$$a = .026S \sqrt{2.8 + .0016S}.$$

And, taking the limiting strain at 80 cwts. per square inch as before, we have:

$$t = \frac{160a}{80 + a},$$

which equation gives the following results:

Span in feet.	Strain in cwts. per sq. in.	Depth.
300	23.8	⅛
400	31	
500	37.7	
600	43.8	
700	48.6	
800	55	
900	60	
1000	64.9	
1100	69.3	

Span in feet.	Strain in cwts. per sq. in.	Depth.
1200	73.5	
1300	77.4	
1400	81	

The value of the multiple $\frac{t}{T-t}$, that is, the weight of iron in terms of the useful load, will be as follows:

Span in feet.			Multiple.
300	$\frac{23.8}{56.2}$	=	.42
400	$\frac{31}{49}$	=	.63
500	$\frac{37.7}{42.3}$	=	.9
600	$\frac{43.8}{36.2}$	=	1.22
700	$\frac{48.6}{31.4}$	=	1.55
800	$\frac{55}{25}$	=	2.2
900	$\frac{60}{20}$	=	3
1000	$\frac{64.9}{15.1}$	=	4.3
1100	$\frac{69.3}{10.7}$	=	6.5
1200	$\frac{73.5}{6.5}$	=	11.3
1300	$\frac{77.4}{2.6}$	=	30
1400		=	∞

Type 9.—*Suspension with Stiffening Girder.*

The combined lightness and strength of an ordinary rope stretched between two supports, and the almost unlimited distance apart at which those points of support might be placed, as compared with that which would have been the limit had the intervening space been spanned by a solid bar even of iron of the same size as the rope, if merely resting on the supports, could hardly have failed to attract, at a very early date, the attention of thoughtful practical men.

When, therefore, the occasion arose of throwing a light structure across a wide obstruction, the similarity of the conditions to the case of the rope with its two distant points of support must almost necessarily have suggested a similar mode of procedure; and knowing the great superiority of the tensional strength of iron over rope, it was only natural that the "suspension bridge," in its simplest form, should be evolved, and that it should be the earliest form of the "long-span bridge," as understood in our definition.

The great instability, however, of this mode of construction was at first inadequately appreciated. It was not until numerous failures had occurred that it began to be clearly understood that a moderate force, applied at regular intervals, would produce an isochronous movement of sufficient extent to effect at last the destruction of the structure. The first modification suggested by these accidents was the insertion of check ties, or even inverted suspension chains, to hold down the platform and so prevent its oscillation. This, however, was a mere empirical remedy, and a more complete and scientific investigation of the conditions of a suspension bridge was required before a successful result could be obtained. As the problem presented no important difficulties, the solution of it was delayed an unreasonable length of time.

A little consideration will show that, to secure the stability of a suspension bridge, it is only necessary to prevent any considerable change of form in the chains, and the consequent tremor and oscillation. This means that the forces acting on the chains should maintain always the same direction and the same relative proportions to one another, although it is not essential to the conditions of equilibrium that the *amount* of these strains should remain constant. This being so, it follows that if the distribution of the load, or other conditions, be such that the relative proportions of the forces would not be maintained, sufficient transverse strength must be provided in the structure to effect the required distribution of the forces acting on the chain. It matters little where this strength be supplied; in some few instances it has been obtained by bracing together two sets of chains, one lying under the other, thus assimilating the chains to an inverted arched rib; in fact, a girder section for the suspension has been patented; ordinarily, however, the required transverse strength is obtained in the construction of the platform, where it is equally efficient, and rather adds to, than detracts from, the elegance of the structure.

Although by this arrangement we effectually counteract the dangerous isochronous movement, still, as the economic depth of the stiffening girder is comparatively very small, and as a double wave of deflection precedes and follows the load as it rolls over the bridge, a considerable amount of vibration must necessarily still exist, although not dangerous in its effects. This objectionable peculiarity of the suspension principle is probably the reason why that class of bridge is almost universally condemned as unfit for railway purposes.

We believe, however, that the bad proportions of the earlier bridges, and the consequent failures, have created an unwarranted prejudice against the system.

There is no theoretical or practical reason why a suspension bridge should not be made of any required degree of rigidity; but whether this could be done *economically* remains to be seen. Whatever can be effected on the principle of the arch may also be obtained in a suspension bridge. Thus, if we were to invert our last type, the arched rib with braced spandrils, jointed at three points, we should obtain a rigid suspension bridge, free to expand and contract under changes of temperature. Again, provided we supply adequate transverse strength to the two halves of the bridge, it is immaterial what form our bracing may assume. As, however, we must have a horizontal girder at the level of the platform, it is more convenient to truss between that member and the arched ribs or chains, as the case may be, than to insert a special member.

It is obvious, then, that it is possible to design an immense variety of forms of rigid suspension bridges; the most elementary type being probably a couple of inclined straight beams, the outer ends of which are attached to the top of the piers, and the near ends jointed together. All that class may be referred to our last type, and the other systems will be included in our remaining types. At present we shall confine our attention to the ordinary suspension bridge with stiffening platform girder.

Chains and Suspending Bars.

In order to obtain great stiffness in the suspending bars and other portions of the structure, and so fit it for its duty of carrying a heavy rolling load with little vibration, we shall provide a mass of metal for the construction of the suspension portion of the bridge, exclusive of the stiffening girder, equal to the amount we have found necessary in an arched bridge carrying an entirely dead load. The strain per square inch will therefore be the same in both instances:

$$t = .021 Sx \sqrt{\frac{y}{x}},$$

where $x = 1.25$, and $y = 3.2 + .002S$.

Lattice Stiffening Girder.

Eliminating the complicating elements the unequal deflections of the chain and girder introduce into the question by assuming a certain amount of preliminary adjustment to be effected during the erection of the bridge, we find the maximum bending moment on the stiffening girder takes place when the bridge is two-thirds loaded, at which time it will, in terms of that on the entire bridge, be equal to $\frac{1}{3}(\frac{2}{3})^2 = \frac{4}{27}$. Therefore, w being the load in cwts. per foot run, and T the strain in cwts. per square inch, the required sectional area of flange will be:

$$a = \frac{S^2 w}{54 d T}.$$

But since all parts of the girder are successively exposed to nearly the same amount of strain, the theoretical mass of the web of the stiffening girder will be about $1\frac{1}{2}$ times that of a similar ordinary girder; therefore economic depth $= \dfrac{Sx}{3y}$. Hence, since weight run of the girder in cwts. (W) equals .03 cwts. × 4 a x, we have:

LONG-SPAN RAILWAY BRIDGES.

$$W = \frac{Swy}{150\,T}.$$

Putting the rolling load $w = 25$ cwts. per foot, the gross useful load = 32 cwts. per foot, and $T = 80$ cwts. per square inch, the weight of the stiffening girder, in terms of the useful load, will be $= \frac{Sy}{15000}$; and the total load on the chains $= 1 + \frac{Sy}{15000}$.

We have the strain in cwts. per square inch on the chains due to the suspension portion of the bridge equal

$$t = .026 S \sqrt{2.56 + .0016 S}.$$

Span in feet.	Strain in cwts. per square in.	Depth of Stiffening Girder.	Deflection of Chains.
300	13.5	$1\frac{1}{8}$	$\frac{1}{5}$
400	18.6		
500	24		
600	29.4		
700	35		
800	40.8		
900	46.8		
1000	53		
1100	59.4		
1200	66		
1300	73		
1400	80	$2\frac{1}{7}$	$\frac{1}{8}$

The total weight of iron in the suspension portion of the bridge and in the stiffening girder will be expressed in terms of the useful load by the equation:

$$\text{Multiple} = \frac{Sy}{15000} + \left\{ \frac{t}{T-t} \left(1 + \frac{Sy}{15000} \right) \right\}.$$

Taking $y = 5 + .0035$, the results are as follows:

Span in feet.	$\frac{Sy}{15000}$	Multiple.
300	.12	.36
400	.17	.53
500	.22	.76
600	.27	1.02
700	.33	1.4
800	.39	1.9
900	.46	2.6
1000	.53	3.5
1100	.61	5.4
1200	.68	8.7
1300	.77	19.4
1400		∞

Type 10.—*Suspended Girder.*

A perfect suspension bridge would be a structure combining the rigidity of a girder with the lightness characteristic of the former system. In attempting to arrive at this desideratum, the question naturally suggests itself, what is the fundamental difference between the two systems, giving them these attributes of rigidity and lightness respectively? Now, the girder is rigid because the depth of bracing, and consequently the resistance offered to change of form, is comparatively large; and the suspension bridge is light, because the compression member of the girder is dispensed with; for although an equivalent resistance must be supplied elsewhere in the land chains and anchorage, yet the mass of metal so employed does not add to the load on the bridge, as it would have done had it been in the form of a compression member.

Let us imagine an inverted bowstring girder, the top compression member being straight, and braced to the curved tension member in the usual manner, then it must be granted that this girder will present equal rigidity with the bowstring type, and that it will also require about the same mass of metal in its construction. But the problem to be solved is how to reduce the mass without impairing the rigidity; and, from what we have already observed, it will be seen that this can only be effected by transferring a portion, or the whole, of the metal required in the top member to some other point where it will be equally efficacious, and will not add to the load on the girder.

Now, the strain on the top member of the girder is compression (+), and it is obvious that if we can put an initial tension (—) on that member, the resulting strain to be provided for will be the difference between those strains, and the mass of metal required will be proportional to that difference. We shall now show how, by a very simple contrivance, we shall be enabled to put any required degree of initial tension on the top member of our girder.

Let one end of the girder be made fast to its pier, and let the other end, instead of resting immediately on the pier, be suspended by an inclined link from it; then, by the resolution of forces, it follows that the initial tension on the top member will bear the same ratio to the entire weight of the bridge as the horizontal component of the inclination of the link bears to double the vertical component. It follows from this that if the inclination of the link be tangential to the curve of equilibrium, due to the load, the initial tension will just neutralize the final compression. As, however, the direction of the tangent to the curve of equilibrium varies with the position of the rolling load, whilst the inclination of the link necessarily remains constant, we shall have to reserve a certain proportion of the compression member to meet the consequent strains. We have, nevertheless, by this arrangement, disposed of the great mass of the compression member without impairing the rigidity or the freedom of movement under changes of temperature, pertaining to the ordinary bowstring girder.

The "suspended girder," as, for want of a better name, we have christened the foregoing type, was, to the best of our belief, first introduced by Mr. Fowler, who proposed constructing his 750 ft. span high-level bridge over the Thames on that principle. There can be no doubt that, had that structure been carried out, the great depth of bracing maintained at the centre of the span—the point of maximum deflection—would have secured for it almost perfect immunity from those vibrating impulses which, although not appearing in theoretical calculations, manifest themselves in a very palpable manner in ordinary "rigid" suspension bridges.

Dead Load.

With a dead load uniformly distributed, the mass of metal required will be similar to that employed in the suspension portion of our last type. The strain in cwts. per square inch due to the weight of the structure itself will therefore be:

$$t = .03\left(.71Sx\sqrt{\frac{\overline{y}}{x}}\right).$$

Rolling Load.

The mass required in the suspension portion will be the same as that for a dead load $= a\left(S + \frac{7d^2}{S}\right)$. That of the bracing will be identical with the corresponding members of the bowstring girder $= a\left(\frac{14d^2}{S} + \frac{S}{5}\right)$. The mass of the compression member necessary to be retained in order to provide for the strains resulting from the unequal distribution of the rolling load, will be $.27aS$. Taking x and y coefficients as before, the total mass for each square inch sectional area of chains will be:

$$1.47Sx + \frac{21d^2y}{S}.$$

But mass × ⅛th span × .03 cwts. equals the moment in cwts.

$$\mu = .03(.18Sx + 2.62d^2y).$$

Now economic $d^2 = \frac{1.47Sx}{21y}$; therefore $d = \frac{10}{38}S\sqrt{\frac{x}{y}}$; and the strain in cwts. per square inch $= \frac{\mu}{d}$, will be:

$$t = .03\left(1.4Sx\sqrt{\frac{\overline{y}}{x}}\right).$$

Mixed Load.

Assuming, as in previous instances, the mixed load to be composed of ¾ rolling and ¼ dead load, the mean coefficient will be $1.4 \times \frac{3}{4} + .71 \times \frac{1}{4} = 1.23$; or taking the coefficient .71 as the unit of measurement, that for the mixed load will be $\frac{1.23}{.71} = 1.73$ times that amount. Hence the strain in cwts. per square inch due to the weight of the structure for carrying a railway will be:

$$t = .021 S x \sqrt{\frac{y}{x}} \left(1.73 - \frac{.73 t}{T} \right).$$

Putting $a = .021 \, Sx \sqrt{\frac{y}{x}}$, when $x = 1.25$ and $y = 3.2 + .002 S$, we have:

$$a = .026 S \sqrt{2.8 + .0016 S}.$$

Taking the limiting strain $T = 80$ cwts. per square inch we have:

$$t = \frac{138 a}{80 + .73 a},$$

which equation gives the following results:

Span in feet.	Strains in cwts. per square inch.	Depth.
300	21	$\frac{1}{7}$
400	27.5	
500	34	
600	40	
700	46	
800	51.5	
900	57	
1000	62	
1100	67	
1200	71.5	
1300	76	
1400	80	$\frac{1}{8}$

The multiple $\dfrac{t}{T-t}$ will have the following values:

Span in feet.			Multiple.
300	$\frac{2\frac{1}{8}}{5}$	=	.36
400	$\frac{5\,7\cdot2}{5\,2\,5}$	=	.52
500	$\frac{3\frac{1}{4}}{8}$	=	.74
600	$\frac{4\,0}{4\,0}$	=	1.00
700	$\frac{4\,6}{3\,4}$	=	1.35
800	$\frac{5\,1\cdot5}{2\,8\cdot5}$	=	1.82
900	$\frac{5\,7}{2\,3}$	=	2.48
1000	$\frac{6\,2}{1\,8}$	=	3.38
1100	$\frac{6\,7}{1\,3}$	=	5.16
1200	$\frac{7\,1\cdot5}{8\cdot5}$	=	8.40
1300	$\frac{7\frac{4}{4}}{4}$	=	19.00
1400		=	∞

Type 11.—*Straight Link Suspension.*

We have found the mass of our last type girder to be similar to that of the bowstring, less a large proportion of the compression member; and in the same manner it can be shown that our present type will be equal to that of the straight-link girder, minus the *whole* of the compression member, together with a certain proportion of the mass of the verticals affected thereby. All our remarks, with reference to the deflection and the conditions of the straight-link girder, will, *mutatis mutandis*, apply equally to the straight-link suspension; consequently, it will be merely necessary here to effect the requisite modifications in our former calculations.

Rolling Load.

The mass of the ties will be the same as before, $= a\left(Sx + \frac{21d^2y}{4S}\right)$. But mass × ⅛ span × .03 cwts. = moment in cwts.; therefore, for 1 square inch,

$$\mu = .03\left(\frac{S^2x}{6} + \frac{7d^2y}{8}\right).$$

Now $\frac{\mu}{d}$ = strain in cwt. per square inch; and since economic $d^2 = \frac{4S^2x}{21y}$, we have $d = \frac{S}{2.3}\sqrt{\frac{x}{y}}$; and the strain

$$t = .03\left(.76Sx\sqrt{\frac{y}{x}}\right).$$

Dead Load.

The mass of the ties will be the same as for the rolling load $= a\left(Sx + \frac{21d^2y}{4S}\right)$. That of the verticals will be the additional amount of $a\left(\frac{13d^2y}{4S}\right)$. The total mass of the girder consequently will be:

$$a\left(Sx + \frac{17d^2y}{2S}\right).$$

But mass × ⅛ span × .03 cwts. = moment in cwts.; hence, if $a = 1$ square inch,

$$\mu = .03\left(\frac{S^2x}{6} + \frac{23d^2y}{16}\right).$$

Since economic $d^2 = \frac{2Sx}{17y}$, we have $d = \frac{S}{2.93}\sqrt{\frac{x}{y}}$; and the strain in cwts. per square inch equal to

$$t = .03\left(.98\ Sx\sqrt{\frac{y}{x}}\right).$$

Mixed Load.

Putting $.03.98\, Sx \left(\sqrt{\dfrac{y}{x}}\right) = a$, we have, when $x = 1.25$, and $y = 3.2 + .002S$, $a = .037S\sqrt{2.56 + .0016S}$; and the strain in cwts. per square inch due to the weight of the straight-link suspension bridge for a railway will be equal to:

$$l = \dfrac{.78aT}{T - .22a}.$$

Taking the limited strain $T = 80$ cwt., we obtain the following results:

Span in feet.	Strain in cwts. per square inch.	Depth.
300	16	$\frac{1}{8}$
400	22	
500	28.6	
600	35.2	
700	43.8	
800	51.8	
900	61.7	
1000	71.6	
1100	82.5	$\frac{1}{7}$

The weight of iron in terms of the useful load will be as follows:

Span in feet.			Multiple.
300	$\frac{16}{64}$	=	.25
400	$\frac{22}{58}$	=	.38
500	$\frac{28.6}{51.4}$	=	.56
600	$\frac{35.2}{44.8}$	=	.78
700	$\frac{43.8}{36.2}$	=	1.22
800	$\frac{51.8}{28.2}$	=	1.85

Span in feet.			Multiple.
900	$\frac{61.7}{18.3}$	=	3.37
1000	$\frac{71.5}{8.5}$	=	7.53
1100		=	∞

Having thus ascertained the weight of iron, in terms of the "useful" load, required in the construction of each of our type girders, the first stage of our inquiry is brought to a termination. It will now be necessary to determine the specific amount of the load in each instance, as upon that depends the weight of the main girders, and then to ascertain the weight of iron required in each bridge, exclusive of that amount; for it is obvious that the respective advantages of the different systems can only be fairly tested by a comparison of the *gross* weight of iron required in the complete construction of the bridge in each instance. We shall not, therefore, waste time in giving a summary of the results in their present incomplete form, but proceed at once with the necessary steps for obtaining a final result.

Now the gross "useful" load on the main girder will be made up of the following elements: 1st, the horizontal bracing necessary to insure the lateral stability of the platform, &c.; 2d, the girders carrying the platform, consisting of cross girders, with longitudinal bearers between each pair, and under each rail; 3d, the platform covering, including the permanent way; 4th, the rolling load. The first two elements will obviously affect the gross weight of iron in the bridge in a double degree, as not only will their own proper weight appear in the sum-total, but it will also affect that amount indirectly, by throwing the additional load on the main girders due to their weight. As the joint weight of the bracing and the platform-girders will be governed partly by the span and partly by the type of the main girders, it will be necessary to estimate the amount for each individual case.

To carry a double line of railway, the weight of the platform-girders will be about 7 cwts. per foot run of the bridge; and this amount will increase in a certain ratio with the span, as the width of platform assumed in the instance of the 300 ft. span would be wholly inadequate to secure lateral stability at the higher spans. Now, the maximum span, to which we shall have occasion to refer, will be about 3200 ft.; and in order to obtain the required degree of lateral stiffness at that span, the effective depth of the horizontal bracing should not be less than 100 ft., and consequently, the span of the cross girder must be at least 100 ft., although possibly no greater length than 25 ft., at the centre, will be occupied by the railway proper. Assuming, then, for the 3200 ft. span, cross-girders 100 ft. span and 30 ft. apart, with longitudinal girders of that span under each rail, the weight of iron in the platform would be as follows:

Total weight of the four girders under rails = $4\frac{3}{4}$ cwt. per foot run of bridge.

Weight of cross girder = $5\frac{1}{2}$ cwt. per foot run.

$$\therefore \frac{5\frac{1}{2} \times 100}{30} = 18\frac{1}{4} \text{ cwts. per foot run of bridge.}$$

The total weight of the platform-girders, will, therefore, be = $4\frac{3}{4} + 18\frac{1}{4} = 23$ cwt. per foot run of the bridge.

The heaviest class of horizontal bracing required in any of our type constructions we shall assume to be equal to $\frac{2}{3}$ths the weight of the preceding platform girders. The total weight of iron required, in addition to that of the main girders, will, therefore, be as follows:

300 ft. span, $1\frac{4}{6} \times 7 = 11$ cwts. per foot span.
3200 " $1\frac{3}{8} \times 23 = 37$ "

Now, 11 : 37 : : $\frac{2}{3}\sqrt{300}$: $\frac{2}{3}\sqrt{3200}$ (nearly); consequently, the weight of iron in the platform girders and the heaviest class of horizontal bracing will be expressed by the simple equation:

Weights in cwts. per foot run of bridge $= \frac{2}{3}\sqrt{\text{Span in ft.}}$

The remainder of the load on a long-span railway bridge may be considered as a constant amount for all spans, equal, for the double line, to 40 cwts. per foot run of the bridge. It follows, therefore, that the "useful" load in cwts. per foot run (L), to be carried by any girder, will be expressed generally by the equation:

$$L = 40 + x\sqrt{S.}$$

For the heaviest description of horizontal bracing, we have found the value of x to be $\frac{2}{3}$. Substituting the proper values for the various required degrees of lateral strength, and classifying our type forms accordingly, we obtain the following equations for the "useful" load:

Types 1, 2, 3, 4.........$L = 40 + \frac{2}{3}\sqrt{S.}$
" 5, 6, 7, 8.........$L = 40 + \frac{7}{12}\sqrt{S.}$
" 9, 10, 11.........$L = 40 + \frac{1}{2}\sqrt{S.}$

Now, W being the *gross* weight of iron in cwts. per foot span required in the construction of a bridge, and M the multiple arrived at in our previous investigations for the particular case under consideration, we have:

$$W = ML + L - 40.$$

Substituting the given values of M and L in this equation, we obtain the results shown in the following tabular form:

Type 1.—*Box Girders with Plate Webs.*

Span in feet.	M.	L—40.	W.
300	.86	11.5	55
400	1.5	13.3	93
500	2.6	15	156
600	5.6	16.3	331
700	25	17.6	1458

Type 2.—*Lattice Girders.*

300	.51	11.5	38
400	.87	13.3	60
500	1.5	15	96
600	2.83	16.3	179
700	8.3	17.8	496

Type 3.—*Bowstring Girders.*

300	.51	11.5	37
400	.79	13.3	56
500	1.19	15	80
600	1.86	16.3	120
700	3	17.6	190
800	5.15	18.8	322
900	15	20	920

Type 4.—*Straight Link and Boom.*

Span in feet.	M.	L.—40.	W.
300	.4	11.5	32
400	.72	13.5	52
500	1.28	15	102
600	2.61	16.3	164
700	9.97	17.6	593

Type 5.—*Cantilever Lattice.*

300	.39	10.1	29
400	.61	11.7	43
500	.88	13.1	59
600	1.23	14.2	80
700	1.7	15.4	109
800	2.5	16.4	157
900	3.6	17.4	225
1000	5.43	18.4	334
1100	9.25	19.3	578
1200	21.	20.3	1469

Type 6.—*Cantilever Lattice, varying depth.*

300	.3	10.1	25
400	.46	11.7	35
500	.65	13.1	47
600	.88	14.2	62
700	1.2	15.4	82
800	1.61	16.4	107
900	2.17	17.4	142
1000	2.96	18.4	191
1100	4.15	19.3	265
1200	6.1	20.3	388
1300	11.2	21.1	705
1400	22	21.8	1380

LONG-SPAN RAILWAY BRIDGES. 79

Type 7.—*Continuous Girder, varying depth.*

Span in feet.	M.	L—40.	W.
300	.3	10.1	25
400	.46	11.7	35
500	.65	13.1	47
600	.85	14.2	60
700	1.1	15.4	76
800	1.33	16.4	91
900	1.61	17.4	110
1000	1.84	18.4	128
1100	2.1	19.3	144
1200	2.48	20.3	169
1300	2.87	21.1	199
1400	3.37	21.8	230
1500	3.91	22.6	267
1600	4.58	23.3	312
1700	5.35	24.1	366
1800	6.47	24.6	449
1900	7.98	25.3	546
2000	10.42	26.1	715
2100	14.32	26.6	981
2200	20	27.3	1375
2300	29	28	2000

Type 8.—*Arched Ribs with Braced Spandrils.*

Span in feet.	M.	L—40.	W.
300	.42	10.1	36
400	.63	11.7	44
500	.9	13.1	60
600	1.22	14.2	80
700	1.55	15.4	100
800	2.2	16.4	135
900	3	17.4	189

Span in feet.	M.	L—40.	W.
1000	4.3	18.4	268
1100	6.5	19.3	408
1200	11.3	20.3	701
1300	30	21.1	1850

Type 9.—*Suspension and Stiffening Girder.*

300	.36	8.6	26
400	.53	10	36
500	.76	11.2	50
600	1.02	12.2	65
700	1.4	13.2	87
800	1.9	14.1	116
900	2.6	15	158
1000	3.5	15.8	211
1100	5.4	16.6	322
1200	8.7	17.3	515
1300	19.4	18	1143.

Type 10.—*Suspended Girders.*

300	.36	8.6	26
400	.52	10	36
500	.74	11.2	49
600	1	12.2	64
700	1.35	13.2	85
800	1.82	14.1	113
900	2.48	15	152
1000	3.38	15.8	204
1100	5.16	16.6	308
1200	8.4	17.3	500
1300	19	18	1120

Type 11.—*Straight Link Suspension.*

Span in feet.	M.	L—40.	W.
300	.5	8.6	21
400	.38	10	30
500	.56	11.2	40
600	.78	12.2	53
700	1.22	13.2	78
800	1.85	14.1	114
900	3.37	15	200
1000	7.53	15.8	436

We have now before us the gross weight of iron per foot span required in the construction of bridges of every class of design, not absolutely eccentric, as we maintain those structures must be which cannot be referred to one or the other of our types.

The weight per foot of the main span will not in itself represent the comparative cost of the superstructure even—much less, as we have shown at the commencement of these papers, that of the entire bridge. With the exception of the first four, all our type constructions are dependent upon extraneous aid for their stability, and the proportional cost of that portion spread over the whole span must obviously be included in the proper weight of the span itself before a comparison can be made with either of the independent girders, where such provision is unnecessary. Thus, given a viaduct of three spans, the centre opening being double the others, then the average weight per foot of the superstructure, if constructed with independent girders, would obviously be the mean weight per foot of the large and small spans; whereas, should some other system, such as the cantilever, be adopted, the average weight per foot would be that due to the long span throughout. We shall defer for the present, however, this branch of our inquiry, and proceed at once with the adaptation of our formulæ to steel structures.

As we do not purpose making a digression on the subject of the comparative strength and other advantages of steel, our paper having already exceeded the proposed limits, we shall have little more to present than the tabular results given by the slightly modified formulæ for similar iron structures.

The general conditions of the several types will be little influenced by the substitution of the new material. The deflections will be proportionally the same, although, of course, the specific amounts will be greater on account of the higher value of the limiting strain. The comparative weights of metal required in the construction of steel and iron long-span bridges will vary in a much higher ratio than the mere ultimate resistances of the two materials. Thus, if the strain due to the weight of the girder itself be 50 cwt. per square inch, there will, with a limiting strain of 80 cwt. per square inch, remain 80—50=30 cwt. per square inch available for the "useful" load; whereas, had the limiting strain been 130 cwt., the residue would have been 130—50=80 cwt. per square inch; consequently in such an instance, although the ultimate resistance of the two materials are only as 13 : 8, the practical available strengths would be as 8 : 3, or about $1\frac{2}{3}$ times the amount which might at first sight have been expected. We shall consider a strain of 130 cwt. per square inch on steel to represent the same factor of safety as a strain of 80 cwt. per square inch on iron. This will be well within the limits, and we take it purposely, so to avoid all exaggeration of the probable advantages accruing to steel as the material for long-span bridges. We shall not make any special modification in our formulæ on account of the slightly increased specific weight of steel, but include all necessary changes in the altered values given to the coefficient y. It will be unnecessary to give the process in the same detail as before; we shall merely give the final formulæ and the combined tables for each type.

Type 1.—*Box Girders with Plate Webs (Steel)*.

As the same causes which make this a disadvantageous form for a long-span *iron* bridge operate with still greater effect when *steel* is the material employed, we do not consider it necessary to investigate the conditions of this type in the present instance.

Type 2.—*Lattice Girders (Steel)*.

The weights of the platform girders and bracing in *steel* we shall take to be $\frac{2}{3}$ of their former amounts, but we shall in all instances add five per cent. for contingencies to the value of W as given by the formulæ. The notation will remain as before; that is, $S=$ span in feet; $T=$ limiting strain $=130$ cwt. per square inch; x and y coefficients depending upon the practical construction of the girders; t the strain in hundred-weights per square inch resulting from the load of the structure itself; M the weight of iron in terms of the "useful" load (L); and W the gross weight of iron per foot run of the entire bridge.

$t = .03\, Sy.$ When $y = 2.7 + .001\, S$, we have:
$t = .105\, S + .00003\, S^2.$

$$M = \frac{t}{T-t}. \qquad L = 40 + \frac{4}{9}\sqrt{S}$$

$$W = ML + L - 40.$$

Span in feet.	t	M.	W.
300	34.2	.35	25
400	46.8	.56	38
500	60	.8	52
600	74.8	1.35	83
700	88.2	2.11	126
800	103.2	3.85	225
900	118.8	10.6	607
1000	135	∞	∞

Type 3.—*Bowstring Girder* (*Steel*).

$$t = \frac{1.44 a T}{T + .44 a}.$$

When $y = 4.2 + .002 S$, we have $a = .04 S \sqrt{3.36 + .0016 S}$

$$M = \frac{t}{T-t}. \qquad L = 40 + \frac{4}{9}\sqrt{S}$$

$$W = ML + L - 40.$$

300	31.	.31	24
400	41.5	.46	33
500	51.5	.65	45
600	61.5	.89	59
700	71	1.2	78

Span in feet.	t	M.	W.
800	80.5	1.62	102
900	90	2.25	140
1000	98	3	184
1100	103.5	3.9	238
1200	115.5	8	480
1300	123	17.5	1047
1400	131	∞	∞

Type 4.—*Straight Link Girder (Steel).*

$$t = \frac{.7aT}{T-.3a}.$$

When $y = 4.2 + .002S$, we have $a = .059 S\sqrt{3.36 + .0016S}$

$$M = \frac{t}{T-t} \qquad L = 40 + \frac{4}{9}\sqrt{S}.$$

$$W = ML + L - 40.$$

300	25.4	.24	21
400	35.8	.38	29
500	46.2	.55	40
600	58.8	.83	55
700	71.5	1.22	78
800	86.5	1.99	123
900	101.6	3.58	214
1000	124.6	23.1	1304
1100	140.7	∞	∞

Type 5.—*Cantilever Lattice, uniform Depth (Steel).*

$t = \sqrt{26S + .008S^2 + 32000} - 179$. When $y = 3.3 + .001S$

$$M = \frac{tb}{8(T-t)}. \qquad L = 40 + \frac{7}{18}\sqrt{S}.$$

$$W = ML + L - 40.$$

8*

Span in feet.	t	M.	W.
300	22	.29	22
400	30	.45	31
500	38	.62	42
600	46	.86	54
700	54	1.13	70
800	63	1.5	91
900	70	1.93	116
1000	78	2.53	152
1100	86	3.34	198
1200	95	4.74	280
1300	103	6.76	397
1400	111	11.1	650
1500	120	22	1288
1600	128	120	7000
1700	137	∞	∞

Type 6.—*Cantilever varying Economic Depth (Steel).*

$$t = \sqrt{21S + .006S^2 + 3200} - 179$$

$$M = \frac{tb}{8(T-t)} \quad L = 40 + \frac{7}{18}\sqrt{S}$$

$$W = ML + L - 40.$$

300	18	.23	19
400	24	.32	24
500	31	.47	33
600	37	.61	41
700	44	.79	48
800	50	1.0	64
900	58	1.27	76
1000	64	1.57	98
1100	71	1.98	124

LONG-SPAN RAILWAY BRIDGES. 87

Span in feet.	t.	M.	W.
1200	79	2.6	158
1300	85	3.23	195
1400	91	4	242
1500	98	5.35	315
1600	106	7.84	481
1700	113	12.0	720
1800	120	22	1326
1900	128	120	7150
2000	134	∞	∞

Type 7.—*Continuous Girder, varying Economic Depth (Steel).*

$$M = \frac{an + 3mb}{a+b} - 1. \quad L = 40 + \frac{7}{18}\sqrt{S}$$

$$W = ML + L - 40.$$

Span in feet.	M.	W.
300	.23	19
400	.32	24
500	.47	33
600	.61	41
700	.79	48
800	.97	63
900	1.15	75
1000	1.3	84
1100	1.5	96
1200	1.7	109
1300	1.91	123
1400	2.12	137
1500	2.37	152
1600	2.63	168
1700	2.94	189

Span in feet.	M.	W.
1800	3.26	210
1900	3.67	237
2000	4.08	264
2100	4.6	297
2200	5.13	332
2300	5.93	385
2400	6.7	434
2500	7.8	506
2600	9	585
2700	11	716
2800	13.2	861
2900	17.5	1141
3000	24.9	1625
3100	52	2831
3200	11.8	7000
4000	∞	∞

Type 8.—*Arched Ribs with Braced Spandrils* (*Steel.*)

$$t = \frac{2a\mathrm{T}}{\mathrm{T}+a}.$$

When $y = 4.5 + 002S$, we have $a = .026S\sqrt{3.6 + 0016S}$

$$M = \frac{t}{\mathrm{T}-t}. \quad L = 40 + \frac{7}{18}\sqrt{S}$$

$$W = ML + L - 40.$$

Span in feet.	t.	M.	W.
300	27.6	.27	21
400	36.8	.39	28
500	44.8	.52	36
600	53.2	.69	46
700	58.5	.87	57
800	68	1.1	71

LONG-SPAN RAILWAY BRIDGES. 89

Span in feet.	t.	M.	W.
900	74.8	1.35	84
1000	81.5	1.68	102
1100	87	2.02	125
1200	94.5	2.66	163
1300	100	3.33	203
1400	105	4.2	255
1500	110	5.5	333
1600	116	8.3	500
1700	120	12.2	735
1800	124	20.7	1240
1900	128	64	3836
2000	133	∞	∞

Type 9.—*Suspension with Stiffening Girder (Steel).*

$$t = .026S\sqrt{3.36 + .0016S}.$$

When $y = 4.2 + .002S$. $L = 40 + \tfrac{1}{3}\sqrt{S}$.

$$M = \frac{Sy_1}{20000} + \left\{ \frac{t}{T-t}\left(1 + \frac{Sy}{20000}\right) \right\}.$$

When $y_1 = 5 + .003S$, $W = ML + L - 40$.

300	15.3	.23	17
400	20.8	.34	24
500	26.5	.45	31
600	32.4	.6	39
700	38.5	.77	49
800	44.8	.97	60
900	51.3	1.22	74
1000	58	1.52	91
1100	64.9	1.9	113
1200	72	2.38	148
1300	79	2.98	175

Span in feet.	t.	M.	W.
1400	87	3.92	228
1500	93	5	289
1600	101	6.98	402
1700	109	10.51	600
1800	117	18.34	1055
1900	125	51.5	2910
2000	134	∞	∞

Type 10.—*Suspended Girder (Steel)*.

$$t = \frac{1.73aT}{T + .73a}.$$

When $y = 4.2 + .002S$, we have $a = .026S\sqrt{3.36 + .0016S}$.

$$M = \frac{t}{T-t}, \quad L = 40 + \tfrac{1}{3}\sqrt{S}$$

$$W = ML + L - 60.$$

300	24.4	.23	17
400	31.6	.32	22
500	39.7	.43	28
600	47.5	.57	36
700	53.7	.72	44
800	64.5	.9	54
900	69	1.13	66
1000	75.5	1.4	81
1100	82.5	1.74	102
1200	88	2.14	124
1300	95	2.7	152
1400	101	3.3	186
1500	105	4.2	234
1600	110	5.5	307
1700	117	9	500

Span in feet.	t.	M.	W.
1800	123	17.5	961
1900	127	42.3	2320
2000	132	∞	∞

Type 11.—*Straight Link Suspension (Steel).*

$$t = \frac{.78 a T}{T - .22 a}.$$

When $y = 4.2 + .002 S$, we have

$$a = .037 S \sqrt{3.36 + .0016 S}$$

$$M = \frac{t}{T-t}, \quad L = 40 + \tfrac{1}{3}\sqrt{S}.$$

$$W = ML + L - 40.$$

300	17.8	.16	14
400	24.2	.23	18
500	31.5	.3	23
600	39.4	.43	30
700	48	.58	38
800	56.5	.77	49
900	64.6	.96	60
1000	73	1.28	78
1100	84.5	1.86	110
1200	95.5	2.77	160
1300	109	5.2	368
1400	118.8	10.6	597
1500	129.5	359	2000
1600			∞

The results of our various investigations are shown collectively in a graphical form in diagrams Nos. 1 and 2, the curved lines of which are obtained by plotting the gross weight of metal in cwts. per foot span, given in the final tables for each type, to the vertical scale of 100 cwts. to the inch. A careful inspection of these diagrams will enable us easily to trace the comparative merits of the respective systems as far as the superstructure of the main span itself is involved, and to note the varying influence of the span in each instance. A glance will be sufficient to assure us that the several types do not maintain the same relative economic positions throughout, since in many instances the lines cross one another, showing that, at the span corresponding to the point of intersection on the diagram, the weight of metal required in the construction of a bridge on either of the systems in question will be identical.

Briefly summarizing the results indicated in the diagram for iron structures, we find that at the span of 300 ft., Type 11—the straight-link suspension bridge—obtains an advantage of some 20 per cent. over any other system, and that it maintains a certain advantage of diminishing value up to 700 ft. span, when it has to resign the lead to Type 7—the continuous girder of varying depth (Sedley's patent)—which type maintains a rapidly increasing advantage over all others up to the limiting span. These two forms of construction, then, within their own proper spheres, appear to be the most economical possible, as far as regards the superstructure of the main span. Of course it is quite possible that in numerous instances anchorage could not be obtained for the suspension bridge, except at a cost which would render even our heaviest type—the box girder—a more economical form of construction.

The system following next in order in the scale of economy is Type 6, the cantilever lattice girder of varying depth, which maintains its relative position throughout, unaffected by the specific length of span. Types 9 and 10, the suspension with stiffening girder, and the suspended girder, succeed the last-named one. Although palpably different both in principle and appearance, the respective weights are almost identical throughout, being, up to 700 ft. span, little different to the preceding type. We now come to Type 5— the cantilever lattice girder of uniform depth—following closely on the heels of the last two systems up to 600 ft. span, when it is superseded by Type 8—the arched rib with braced spandrils. The independent girders, as might fairly be expected, occupy the lowest place on the list, although at 300 ft. span Type 4— the straight-link girder—shows a slight advantage over the arch. Within the limits of 400 or 500 ft. span, the straight link is the most economic form of the independent girder; above that span the bowstring girder surpasses it. Types 2 and 1, the lattice and box girders, conclude the list.

We have already insisted that the order of economy thus exhibited holds good only with reference to the superstructure of the main span, and that it does not represent the comparative costs of complete structures on the different systems, for which, indeed, no general rule could possibly be given. Some of our types require loftier and, *cæteris paribus*, more expensive piers than others; thus the piers for a suspension bridge will be higher than those for an arch, and the land-chains required in the former system will be another cause of excess. Probably the fairest way of arriving approximately at the true relative economy of the different systems in ordinary cases will be by ascertaining the gross *average* weight of iron per foot run required in the construction of viaducts consisting of three spans, of which the side spans are one-half the opening of the centre one. Under these circumstances, with the exception of the independent girders, the average weight per foot run will be that due to the long span, whilst in those instances the smaller weight per foot of the side span will reduce the *average* weight per foot of the entire structure, as may be seen by the comparison of the following tables with former ones :—

$S =$ span of centre opening; $\dfrac{S}{2} =$ span of side openings.

$W =$ average weight of iron in cwts. per foot run of bridge.

Iron Girder Viaducts.

S.	Box. W.	Lattice. W.	Bowstring. W.	Straight link. W.
300	44	32	31	27
400	66	45	43	39
500	102	65	57	65
600	193	109	79	98
700	766	273	118	318
800			189	
900			494	

Steel Girder Viaducts.

S.	Lattice. W.	Bowstring. W.	Straight link. W.
300	21	21	18
400	29	26	23
500	37	34	30
600	54	42	38
700	79	54	52
800	132	68	76
900	326	90	125
1000		115	672
1100		145	
1200		270	
1300		558	

From the preceding tables it is at once apparent that, at the shorter spans, the difference of weight between the independent girders and those dependent upon some extraneous means of support is much less marked than before. Indeed, in some instances the conditions are reversed; thus, at 300 ft. span both the lattice and bowstring girders show a certain positive advantage over the arch, which the more economical system of the straight link retains up to 400 ft. span even.

The positions of the several types in the scale of economy would, obviously, be again shifted if the side spans were taken at a different ratio to the main span than ¼; but we believe it is unnecessary on that account to extend our investigations.

Of the numerous practical considerations and contingencies to be duly weighed and carefully estimated, before the fitness of a design for a long-span railway bridge could be satisfactorily determined, none are more important than those affecting the facility of erection. In the majority of instances scaffolding would impose a most extravagant charge on the finished structure for mere temporary works, even if the adoption of it were not practically prohibited by the necessity of maintaining the navigable channel—across which, probably, the bridge has to be thrown—free from all obstructions.

Under such circumstances there are obviously but two alternatives: the bridge must either be built at the nearest available spot, and lifted or slid bodily into place; or the design must be such that the structure may be built *in situ*, without adventitious support.

The latter alternative is incapable of application to either of our first four types—the independent girders. The method adopted in the instance of the Britannia bridge, where the tubes were built on shore, floated, and lifted into place, is an example of the former alternative applied to a box girder; and Schwedler's six-span bridge, each span of which was conveyed in a set of six-wheel wagons to the required spot, and lifted bodily into the place by four hydraulic presses, illustrates the adaptation of the same system to a bowstring bridge.

The more convenient process of constructing the bridge in the position it is permanently to hold may, in one form or another, be employed for any of our type forms, not being one of the first four. It was proposed to build Mr. Fowler's Severn bridge in successive bays projecting from each side of the two main piers, carrying on the process till the two opposing centre halves met, and formed a continuous structure. The large cantilevers of Sedley's bridges are all built in a similar manner, and the small centre girder is lifted or slid into place.

In an arched bridge, to be erected without scaffolding, temporary ties in some form are indispensable. If it is to be lifted into place, the feet of the ribs must be tied together; and if it is to be built in position, the arched rib must be tied back to the abutments, as proposed in the first design for the Britannia bridge.

Types 9, 10, and 11 might all be dealt with in a manner analogous to that by which the Ohio suspension bridge was successfully erected. Wire ropes were paid out over the stern of a vessel, and laid on the bed of the river, as if they were electric cables. At convenient times they were smartly hoisted up to the summits of the towers, and a slight platform placed on them to facilitate the construction of the main cables.

If due consideration be given to these and other special conditions affecting each individual case, we think, with the assistance of the tables already given, there will be little difficulty in ascertaining with a considerable amount of accuracy the most suitable form of construction possible for any " *long-span railway bridge.*"

Type 7a. Sedley's Patent. (Surut Bridge)

Type 7 Continuous Girder, varying depth.

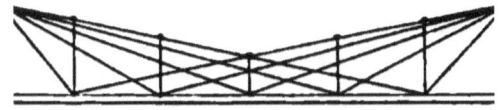

Type 11 Straight Link Suspension.

INDEX.

A.

	PAGE
Arched ribs with braced sprandrils	53
Arched ribs with braced spandrils, dead load on	56
Arched ribs with braced spandrils, mixed load on	59
Arched ribs with braced spandrils, rolling load on	57

B.

Bollman truss	33
Bowstring girders	25
Bowstring girders, dead load on	28
Bowstring girders, mixed load on	31
Bowstring girders, rolling load on	29
Box girders with plate webs (steel)	83
Bridges, long-span railway	9
Britannia bridge	96

C.

Cantilever girders uniform depth	41
Cantilever lattice girders, varying economic depth	45
Chains and suspending bars	63
Continuous girder	49

G.

Girders	14

I.

Iron girder viaducts	94

L.

	PAGE
Lattice girder	23
Lattice girders (steel)	83
Lattice stiffening girder	64
Load for arched ribs with braced spandrils	79
Load for bowstring girders	77
Load for box girders with plate webs	77
Load for cantilever lattice	78
Load for cantilever lattice, varying depth	78
Load for continuous girder	79
Load for lattice girders	77
Load for straight-link and boom	78
Load for straight-link suspension	81
Load for suspended girders	80
Load for suspension and stiffening girder	80

O.

Ohio suspension bridge	97

R.

Railway bridges, long-span	9

S.

Sedley's patent	49
Schwedler's bridge	96
Steel, bridges of	83
Steel girder viaducts	95
Straight-link girders	33
Straight-link girders, dead load on	36
Straight-link girders, mixed load on	39
Straight-link girders, rolling load on	38
Straight-link suspension	71
" " " dead load	72
" " " mixed load	73
Summary of results for iron structures	92
Suspended girder	66
Suspended girder, dead load	68
" " rolling load	69
" " mixed load	69
Suspension with stiffening girder	60

V.

	PAGE
Viaducts	94
Viaducts, steel girder	95

W.

Weight borne by bridges	77
Weight for arched ribs with braced spandrils (steel)	88
Weight for bowstring girders (steel)	84
Weight for cantilever lattice, uniform depth (steel)	85
Weight for cantilever varying economic depth (steel)	86
Weight for continuous girder varying economic depth (steel)	87
Weight for lattice girders (steel)	84
Weight for straight-link girder (steel)	85
Weight for straight-link suspension (steel)	91
Weight for suspended girder (steel)	90
Weight for suspension with stiffening girder (steel)	89
Weight of iron for girders	14
Weight of viaducts	94

THE END.

CATALOGUE
OF
PRACTICAL AND SCIENTIFIC BOOKS,
PUBLISHED BY
HENRY CAREY BAIRD,
INDUSTRIAL PUBLISHER,
No. 406 WALNUT STREET,
PHILADELPHIA.

☞ Any of the Books comprised in this Catalogue will be sent by mail, free of postage, at the publication price.

☞ This Catalogue will be sent, free of postage, to any one who will furnish the publisher with his address.

ARMENGAUD, AMOUROUX, AND JOHNSON.—THE PRACTICAL DRAUGHTSMAN'S BOOK OF INDUSTRIAL DESIGN, AND MACHINIST'S AND ENGINEER'S DRAWING COMPANION: Forming a complete course of Mechanical Engineering and Architectural Drawing. From the French of M. Armengaud the elder, Prof. of Design in the Conservatoire of Arts and Industry, Paris, and MM. Armengaud the younger and Amouroux, Civil Engineers. Rewritten and arranged, with additional matter and plates, selections from and examples of the most useful and generally employed mechanism of the day. By WILLIAM JOHNSON, Assoc. Inst. C. E., Editor of "The Practical Mechanic's Journal." Illustrated by 50 folio steel plates and 50 wood-cuts. A new edition, 4to. . $10 00

ARROWSMITH.—PAPER-HANGER'S COMPANION: A Treatise in which the Practical Operations of the Trade are Systematically laid down: with Copious Directions Preparatory to Papering; Preventives against the Effect of Damp on Walls; the Various Cements and Pastes adapted to the Several Purposes of the Trade; Observations and Directions for the Panelling and Ornamenting of Rooms, &c. By JAMES ARROWSMITH, Author of "Analysis of Drapery," &c. 12mo, cloth $1 25

BULLOCK.— THE RUDIMENTS OF ARCHITECTURE AND BUILDING:
For the use of Architects, Builders, Draughtsmen, Machinists, Engineers, and Mechanics. Edited by JOHN BULLOCK, author of "The American Cottage Builder." Illustrated by 250 engravings. In one volume 8vo. . . . $3 50

BURGH.—PRACTICAL ILLUSTRATIONS OF LAND AND MARINE ENGINES:
Showing in detail the Modern Improvements of High and Low Pressure, Surface Condensation, and Super-heating, together with Land and Marine Boilers. By N. P. BURGH, Engineer. Illustrated by twenty plates, double elephant folio, with text. $21 00

BURGH.—PRACTICAL RULES FOR THE PROPORTIONS OF MODERN ENGINES AND BOILERS FOR LAND AND MARINE PURPOSES.
By N. P. BURGH, Engineer. 12mo. . . . $2 00

BURGH.—THE SLIDE-VALVE PRACTICALLY CONSIDERED:
By N. P. BURGH, author of "A Treatise on Sugar Machinery," "Practical Illustrations of Land and Marine Engines," "A Pocket-Book of Practical Rules for Designing Land and Marine Engines, Boilers," etc. etc. etc. Completely illustrated. 12mo. $2 00

BYRN.—THE COMPLETE PRACTICAL BREWER:
Or, Plain, Accurate, and Thorough Instructions in the Art of Brewing Beer, Ale, Porter, including the Process of making Bavarian Beer, all the Small Beers, such as Root-beer, Ginger-pop, Sarsaparilla-beer, Mead, Spruce beer, etc. etc. Adapted to the use of Public Brewers and Private Families. By M. LA FAYETTE BYRN, M. D. With illustrations. 12mo. $1 25

BYRN.—THE COMPLETE PRACTICAL DISTILLER:
Comprising the most perfect and exact Theoretical and Practical Description of the Art of Distillation and Rectification; including all of the most recent improvements in distilling apparatus; instructions for preparing spirits from the numerous vegetables, fruits, etc.; directions for the distillation and preparation of all kinds of brandies and other spirits, spirituous and other compounds, etc. etc.; all of which is so simplified that it is adapted not only to the use of extensive distillers, but for every farmer, or others who may wish to engage in the art of distilling By M. LA FAYETTE BYRN, M. D. With numerous engravings. In one volume, 12mo. $1 50

BYRNE.—POCKET BOOK FOR RAILROAD AND CIVIL ENGINEERS:
Containing New, Exact, and Concise Methods for Laying out Railroad Curves, Switches, Frog Angles and Crossings; the Staking out of work; Levelling; the Calculation of Cuttings; Embankments; Earth-work, etc. By OLIVER BYRNE. Illustrated, 18mo., full bound $1 75

BYRNE.—THE HANDBOOK FOR THE ARTISAN, MECHANIC, AND ENGINEER:
By OLIVER BYRNE. Illustrated by 185 Wood Engravings. 8vo.
$5 00

BYRNE.—THE ESSENTIAL ELEMENTS OF PRACTICAL MECHANICS:
For Engineering Students, based on the Principle of Work. By OLIVER BYRNE. Illustrated by Numerous Wood Engravings, 12mo. $3 63

BYRNE.—THE PRACTICAL METAL-WORKER'S ASSISTANT:
Comprising Metallurgic Chemistry; the Arts of Working all Metals and Alloys; Forging of Iron and Steel; Hardening and Tempering; Melting and Mixing; Casting and Founding; Works in Sheet Metal; the Processes Dependent on the Ductility of the Metals; Soldering; and the most Improved Processes and Tools employed by Metal-Workers. With the Application of the Art of Electro-Metallurgy to Manufacturing Processes; collected from Original Sources, and from the Works of Holtzapffel, Bergeron, Leupold, Plumier, Napier, and others. By OLIVER BYRNE. A New, Revised, and improved Edition, with Additions by John Scoffern, M. B , William Clay, Wm. Fairbairn, F. R. S., and James Napier. With Five Hundred and Ninety-two Engravings; Illustrating every Branch of the Subject. In one volume, 8vo. 652 pages . $7 00

BYRNE.—THE PRACTICAL MODEL CALCULATOR:
For the Engineer, Mechanic, Manufacturer of Engine Work, Naval Architect, Miner, and Millwright. By OLIVER BYRNE. 1 volume, 8vo., nearly 600 pages $4 50

BEMROSE.—MANUAL OF WOOD CARVING: With Practical Illustrations for Learners of the Art, and Original and Selected designs. By WILLIAM BEMROSE, Jr. With an Introduction by LLEWELLYN JEWITT, F. S. A., etc. With 128 Illustrations. 4to., cloth $3 00

BAIRD.—PROTECTION OF HOME LABOR AND HOME PRO-
DUCTIONS NECESSARY TO THE PROSPERITY OF THE
AMERICAN FARMER:
By HENRY CAREY BAIRD. 8vo., paper 10

BAIRD.—STANDARD WAGES COMPUTING TABLES:
An Improvement in all former Methods of Computation, so arranged that wages for days, hours, or fractions of hours, at a specified rate per day or hour, may be ascertained at a glance. By T. SPANGLER BAIRD. Oblong folio $5 00

BISHOP.—A HISTORY OF AMERICAN MANUFACTURES:
From 1608 to 1866: exhibiting the Origin and Growth of the Principal Mechanic Arts and Manufactures, from the Earliest Colonial Period to the Present Time; with a Notice of the Important Inventions, Tariffs, and the Results of each Decennial Census. By J. LEANDER BISHOP, M. D.; to which are added Notes on the Principal Manufacturing Centres and Remarkable Manufactories. By EDWARD YOUNG and EDWIN T. FREEDLEY. In three vols. 8vo. $10 00

BOX.—A PRACTICAL TREATISE ON HEAT AS APPLIED TO
THE USEFUL ARTS:
For the use of Engineers, Architects, etc. By THOMAS BOX, author of "Practical Hydraulics." Illustrated by 14 plates, containing 114 figures. 12mo. $4 25

CABINET MAKER'S ALBUM OF FURNITURE:
Comprising a Collection of Designs for the Newest and Most Elegant Styles of Furniture. Illustrated by Forty-eight Large and Beautifully Engraved Plates. In one volume, oblong
$5 00

CHAPMAN.—A TREATISE ON ROPE-MAKING:
As practised in private and public Rope-yards, with a Description of the Manufacture, Rules, Tables of Weights, etc., adapted to the Trade; Shipping, Mining, Railways, Builders, etc. By ROBERT CHAPMAN. 24mo. $1 50

CRAIK.—THE PRACTICAL AMERICAN MILLWRIGHT AND
MILLER.
By DAVID CRAIK, Millwright. Illustrated by numerous wood engravings, and two folding plates. 8vo. (*In press.*)

CALVERT.—LECTURES ON COAL-TAR COLORS, AND ON RECENT IMPROVEMENTS AND PROGRESS IN DYEING AND CALICO PRINTING:

Embodying Copious Notes taken at the last London International Exhibition, and *Illustrated with Numerous Patterns of Aniline and other Colors.* By F. Grace Calvert, F. R. S., F. C. S., Professor of Chemistry at the Royal Institution, Manchester, Corresponding Member of the Royal Academies of Turin and Rouen; of the Pharmaceutical Society of Paris; Société Industrielle de Mulhouse, etc. In one volume, 8vo., cloth $1 50

CAMPIN.—A PRACTICAL TREATISE ON MECHANICAL ENGINEERING:

Comprising Metallurgy, Moulding, Casting, Forging, Tools, Workshop Machinery, Mechanical Manipulation, Manufacture of Steam-engines, etc. etc. With an Appendix on the Analysis of Iron and Iron Ores. By Francis Campin, C. E. To which are added, Observations on the Construction of Steam Boilers, and Remarks upon Furnaces used for Smoke Prevention; with a Chapter on Explosions. By R. Armstrong, C. E., and John Bourne. Rules for Calculating the Change Wheels for Screws on a Turning Lathe, and for a Wheel-cutting Machine. By J. La Nicca. Management of Steel, including Forging, Hardening, Tempering, Annealing, Shrinking, and Expansion. And the Case-hardening of Iron. By G. Ede. 8vo. Illustrated with 29 plates and 100 wood engravings.
$6 00

CAMPIN.—THE PRACTICE OF HAND-TURNING IN WOOD, IVORY, SHELL, ETC.:

With Instructions for Turning such works in Metal as may be required in the Practice of Turning Wood, Ivory, etc. Also an Appendix on Ornamental Turning. By Francis Campin, with Numerous Illustrations, 12mo., cloth . . $3 00

CAPRON DE DOLE.—DUSSAUCE.—BLUES AND CARMINES OF INDIGO.

A Practical Treatise on the Fabrication of every Commercial Product derived from Indigo. By Felicien Capron de Dole. Translated, with important additions, by Professor H. Dussauce. 12mo. $2 50

CAREY.—THE WORKS OF HENRY C. CAREY:

CONTRACTION OR EXPANSION? REPUDIATION OR RESUMPTION? Letters to Hon. Hugh McCulloch. 8vo. 38

FINANCIAL CRISES, their Causes and Effects. 8vo. paper 25

HARMONY OF INTERESTS; Agricultural, Manufacturing, and Commercial. 8vo., paper $1 00
    Do.   do.   cloth . . . $1 50

LETTERS TO THE PRESIDENT OF THE UNITED STATES. Paper $1 00

MANUAL OF SOCIAL SCIENCE. Condensed from Carey's "Principles of Social Science." By KATE McKEAN. 1 vol. 12mo. $2 25

MISCELLANEOUS WORKS: comprising "Harmony of Interests," "Money," "Letters to the President," "French and American Tariffs," "Financial Crises," "The Way to Outdo England without Fighting Her," "Resources of the Union," "The Public Debt," "Contraction or Expansion," "Review of the Decade 1857—'67," "Reconstruction," etc. etc. 1 vol. 8vo., cloth $1 50

MONEY: A LECTURE before the N. Y. Geographical and Statistical Society. 8vo., paper 25

PAST, PRESENT, AND FUTURE. 8vo. . . . $2 50

PRINCIPLES OF SOCIAL SCIENCE. 3 volumes 8vo., cloth $10 00

REVIEW OF THE DECADE 1857—'67. 8vo., paper 50

RECONSTRUCTION: INDUSTRIAL, FINANCIAL, AND POLITICAL. Letters to the Hon. Henry Wilson, U. S. S. 8vo paper 50

THE PUBLIC DEBT, LOCAL AND NATIONAL. How to provide for its discharge while lessening the burden of Taxation. Letter to David A. Wells, Esq., U. S. Revenue Commission. 8vo., paper 25

THE RESOURCES OF THE UNION. A Lecture read, Dec. 1865, before the American Geographical and Statistical Society, N. Y., and before the American Association for the Advancement of Social Science, Boston . . . 50

THE SLAVE TRADE, DOMESTIC AND FOREIGN; Why it Exists, and How it may be Extinguished. 12mo., cloth $1 50

THE WAY TO OUTDO ENGLAND WITHOUT FIGHTING HER. Letters to the Hon. Schuyler Colfax. 8vo., paper $1 00

CAMUS.—A TREATISE ON THE TEETH OF WHEELS:
Demonstrating the best forms which can be given to them for the purposes of Machinery, such as Mill-work and Clock-work. Translated from the French of M. Camus. By John I. Hawkins. Illustrated by 40 plates. 8vo. $3 00

CLOUGH.—THE CONTRACTOR'S MANUAL AND BUILDER'S PRICE-BOOK:
Designed to elucidate the method of ascertaining, correctly, the value and Quantity of every description of Work and Materials used in the Art of Building, from their Prime Cost in any part of the United States, collected from extensive experience and observation in Building and Designing; to which are added a large variety of Tables, Memoranda, etc., indispensable to all engaged or concerned in erecting buildings of any kind. By A. B. Clough, Architect, 24mo., cloth   75

COLBURN.—THE GAS-WORKS OF LONDON:
Comprising a sketch of the Gas-works of the city, Process of Manufacture, Quantity Produced, Cost, Profit, etc. By Zerah Colburn. 8vo., cloth 75

COLBURN.—THE LOCOMOTIVE ENGINE:
Including a Description of its Structure, Rules for Estimating its Capabilities, and Practical Observations on its Construction and Management. By Zerah Colburn. Illustrated. A new edition. 12mo. $1 25

COLBURN AND MAW.—THE WATER-WORKS OF LONDON:
Together with a Series of Articles on various other Water-works. By Zerah Colburn and W. Maw. Reprinted from "Engineering." In one volume, 8vo. . . . $4 00

DAGUERREOTYPIST AND PHOTOGRAPHER'S COMPANION:
12mo., cloth $1 25

DUPLAIS.—A COMPLETE TREATISE ON THE DISTILLATION AND PREPARATION OF ALCOHOLIC AND OTHER LIQUORS:
From the French of M. Duplais. Translated and Edited by M. McKennie, M. D. Illustrated. 8vo. (*In press.*)

DIRCKS.—PERPETUAL MOTION:
Or Search for Self-Motive Power during the 17th, 18th, and 19th centuries. Illustrated from various authentic sources in Papers, Essays, Letters, Paragraphs, and numerous Patent Specifications, with an Introductory Essay by Henry Dircks, C. E. Illustrated by numerous engravings of machines. 12mo., cloth $3 50

DIXON.—THE PRACTICAL MILLWRIGHT'S AND ENGINEER'S GUIDE:
Or Tables for Finding the Diameter and Power of Cogwheels; Diameter, Weight, and Power of Shafts; Diameter and Strength of Bolts, etc. etc. By Thomas Dixon. 12mo., cloth. $1 50

DUNCAN.—PRACTICAL SURVEYOR'S GUIDE:
Containing the necessary information to make any person, of common capacity, a finished land surveyor without the aid of a teacher. By Andrew Duncan. Illustrated. 12mo., cloth. $1 25

DUSSAUCE.—A NEW AND COMPLETE TREATISE ON THE ARTS OF TANNING, CURRYING, AND LEATHER DRESSING:
Comprising all the Discoveries and Improvements made in France, Great Britain, and the United States. Edited from Notes and Documents of Messrs. Sallerou, Grouvelle, Duval, Dessables, Labarraque, Payen, René, De Fontenelle, Malapeyre, etc. etc. By Prof. H. Dussauce, Chemist. Illustrated by 212 wood engravings. 8vo. $10 00

DUSSAUCE.—A GENERAL TREATISE ON THE MANUFACTURE OF SOAP, THEORETICAL AND PRACTICAL:
Comprising the Chemistry of the Art, a Description of all the Raw Materials and their Uses. Directions for the Establishment of a Soap Factory, with the necessary Apparatus, Instructions in the Manufacture of every variety of Soap, the Assay and Determination of the Value of Alkalies, Fatty Substances, Soaps, etc. etc. By Professor H. Dussauce. With an Appendix, containing Extracts from the Reports of the International Jury on Soaps, as exhibited in the Paris Universal Exposition, 1867, numerous Tables, etc. etc. Illustrated by engravings. In one volume 8vo. of over 800 pages $10 00

DUSSAUCE.—A PRACTICAL GUIDE FOR THE PERFUMER:
Being a New Treatise on Perfumery the most favorable to the Beauty without being injurious to the Health, comprising a Description of the substances used in Perfumery, the Formulæ of more than one thousand Preparations, such as Cosmetics, Perfumed Oils, Tooth Powders, Waters, Extracts, Tinctures, Infusions, Vinaigres, Essential Oils, Pastels, Creams, Soaps, and many new Hygienic Products not hitherto described. Edited from Notes and Documents of Messrs. Debay, Lunel, etc. With additions by Professor H. DUSSAUCE, Chemist. 12mo. $3 00

DUSSAUCE.—PRACTICAL TREATISE ON THE FABRICATION OF MATCHES, GUN COTTON, AND FULMINATING POWDERS.
By Professor H. DUSSAUCE. 12mo. . . . $3 00

DUSSAUCE.—A GENERAL TREATISE ON THE MANUFACTURE OF VINEGAR, THEORETICAL AND PRACTICAL.
Comprising the various methods, by the slow and the quick processes, with Alcohol, Wine, Grain, Cider, and Molasses, as well as the Fabrication of Wood Vinegar, etc. By Prof. H. DUSSAUCE. 12mo. (*In press.*)

DE GRAFF.—THE GEOMETRICAL STAIR-BUILDERS' GUIDE:
Being a Plain Practical System of Hand-Railing, embracing all its necessary Details, and Geometrically Illustrated by 22 Steel Engravings; together with the use of the most approved principles of Practical Geometry. By SIMON DE GRAFF, Architect. 4to. $5 00

DYER AND COLOR-MAKER'S COMPANION:
Containing upwards of two hundred Receipts for making Colors, on the most approved principles, for all the various styles and fabrics now in existence; with the Scouring Process, and plain Directions for Preparing, Washing-off, and Finishing the Goods. In one vol. 12mo. $1 25

EASTON.—A PRACTICAL TREATISE ON STREET OR HORSE-POWER RAILWAYS:
Their Location, Construction, and Management; with General Plans and Rules for their Organization and Operation; together with Examinations as to their Comparative Advantages over the Omnibus System, and Inquiries as to their Value for Investment; including Copies of Municipal Ordinances relating thereto. By ALEXANDER EASTON, C. E. Illustrated by 23 plates, 8vo., cloth $2 00

FORSYTH.—BOOK OF DESIGNS FOR HEAD-STONES, MURAL, AND OTHER MONUMENTS:
Containing 78 Elaborate and Exquisite Designs. By FORSYTH. 4to. (*In press*)

FAIRBAIRN.—THE PRINCIPLES OF MECHANISM AND MACHINERY OF TRANSMISSION:
Comprising the Principles of Mechanism, Wheels, and Pulleys, Strength and Proportions of Shafts, Couplings of Shafts, and Engaging and Disengaging Gear. By WILLIAM FAIRBAIRN, Esq., C. E., LL. D., F. R. S., F. G. S., Corresponding Member of the National Institute of France, and of the Royal Academy of Turin; Chevalier of the Legion of Honor, etc. etc. Beautifully illustrated by over 150 wood-cuts. In one volume 12mo. $2 50

FAIRBAIRN.—PRIME-MOVERS:
Comprising the Accumulation of Water-power; the Construction of Water-wheels and Turbines; the Properties of Steam; the Varieties of Steam-engines and Boilers and Wind-mills. By WILLIAM FAIRBAIRN, C. E., LL. D., F. R. S., F. G. S. Author of "Principles of Mechanism and the Machinery of Transmission." With Numerous Illustrations. In one volume. (In press.)

FLAMM.—A PRACTICAL GUIDE TO THE CONSTRUCTION OF ECONOMICAL HEATING APPLICATIONS FOR SOLID AND GASEOUS FUELS:
With the Application of Concentrated Heat, and on Waste Heat, for the Use of Engineers, Architects, Stove and Furnace Makers, Manufacturers of Fire Brick, Zinc, Porcelain, Glass, Earthenware, Steel, Chemical Products, Sugar Refiners, Metallurgists, and all others employing Heat. By M. PIERRE FLAMM, Manufacturer. Illustrated. Translated from the French. One volume, 12mo. (In press.)

GILBART.—A PRACTICAL TREATISE ON BANKING:
By JAMES WILLIAM GILBART. To which is added: THE NATIONAL BANK ACT AS NOW (1868) IN FORCE. 8vo. $4 50

GOTHIC ALBUM FOR CABINET MAKERS:
Comprising a Collection of Designs for Gothic Furniture. Illustrated by twenty-three large and beautifully engraved plates. Oblong $3 00

GRANT.—BEET-ROOT SUGAR AND CULTIVATION OF THE BEET:
By E. B. Grant. 12mo. $1 25

GREGORY.—MATHEMATICS FOR PRACTICAL MEN:
Adapted to the Pursuits of Surveyors, Architects, Mechanics, and Civil Engineers. By Olinthus Gregory. 8vo., plates, cloth $3 00

GRISWOLD.—RAILROAD ENGINEER'S POCKET COMPANION.
Comprising Rules for Calculating Deflection Distances and Angles, Tangential Distances and Angles, and all Necessary Tables for Engineers; also the art of Levelling from Preliminary Survey to the Construction of Railroads, intended Expressly for the Young Engineer, together with Numerous Valuable Rules and Examples. By W. Griswold. 12mo., tucks. $1 75

GUETTIER.—METALLIC ALLOYS:
Being a Practical Guide to their Chemical and Physical Properties, their Preparation, Composition, and Uses. Translated from the French of A. Guettier, Engineer and Director of Founderies, author of "La Fouderie en France," etc. etc. By A. A. Fesquet, Chemist and Engineer. In one volume, 12mo. (In press, *shortly to be published.*)

HATS AND FELTING:
A Practical Treatise on their Manufacture. By a Practical Hatter. Illustrated by Drawings of Machinery, &c., 8vo. $1 25

HAY.—THE INTERIOR DECORATOR:
The Laws of Harmonious Coloring adapted to Interior Decorations: with a Practical Treatise on House-Painting. By D. R. Hay, House-Painter and Decorator. Illustrated by a Diagram of the Primary, Secondary, and Tertiary Colors. 12mo. $2 25

HUGHES.—AMERICAN MILLER AND MILLWRIGHT'S ASSISTANT:
By Wm. Carter Hughes. A new edition. In one volume, 12mo. $1 50

HUNT.—THE PRACTICE OF PHOTOGRAPHY.
By ROBERT HUNT, Vice-President of the Photographic Society, London, with numerous illustrations. 12mo., cloth . 75

HURST.—A HAND-BOOK FOR ARCHITECTURAL SURVEYORS:
Comprising Formulœ useful in Designing Builder's work, Table of Weights, of the materials used in Building, Memoranda connected with Builders' work, Mensuration, the Practice of Builders' Measurement, Contracts of Labor, Valuation of Property, Summary of the Practice in Dilapidation, etc. etc. By J. F. HURST, C. E. 2d edition, pocket-book form, full bound $2 50

JERVIS.—RAILWAY PROPERTY:
A Treatise on the Construction and Management of Railways; designed to afford useful knowledge, in the popular style, to the holders of this class of property; as well as Railway Managers, Officers, and Agents. By JOHN B. JERVIS, late Chief Engineer of the Hudson River Railroad, Croton Aqueduct, &c. One vol. 12mo., cloth $2 00

JOHNSON.—A REPORT TO THE NAVY DEPARTMENT OF THE UNITED STATES ON AMERICAN COALS:
Applicable to Steam Navigation and to other purposes. By WALTER R. JOHNSON. With numerous illustrations. 607 pp. 8vo., half morocco $6 00

JOHNSON.—THE COAL TRADE OF BRITISH AMERICA:
With Researches on the Characters and Practical Values of American and Foreign Coals. By WALTER R. JOHNSON, Civil and Mining Engineer and Chemist. 8vo. . . . $2 00

JOHNSTON.—INSTRUCTIONS FOR THE ANALYSIS OF SOILS, LIMESTONES, AND MANURES.
By J. W. F. JOHNSTON. 12mo. 35

KEENE.—A HAND-BOOK OF PRACTICAL GAUGING,
For the Use of Beginners, to which is added A Chapter on Distillation, describing the process in operation at the Custom House for ascertaining the strength of wines. By JAMES B. KEENE, of H. M. Customs. 8vo. $1 25

KENTISH.—A TREATISE ON A BOX OF INSTRUMENTS,
And the Slide Rule; with the Theory of Trigonometry and Logarithms, including Practical Geometry, Surveying, Measuring of Timber, Cask and Malt Gauging, Heights, and Distances. By THOMAS KENTISH. In one volume. 12mo. . $1 25

KOBELL.—ERNI.—MINERALOGY SIMPLIFIED:
A short method of Determining and Classifying Minerals, by means of simple Chemical Experiments in the Wet Way. Translated from the last German Edition of F. VON KOBELL, with an Introduction to Blowpipe Analysis and other additions. By HENRI ERNI, M. D., Chief Chemist, Department of Agriculture, author of "Coal Oil and Petroleum." In one volume, 12mo. $2 50

LAFFINEUR.—A PRACTICAL GUIDE TO HYDRAULICS FOR TOWN AND COUNTRY;
Or a Complete Treatise on the Building of Conduits for Water for Cities, Towns, Farms, Country Residences, Workshops, etc. Comprising the means necessary for obtaining at all times abundant supplies of Drinkable Water. Translated from the French of M. JULES LAFFINEUR, C. E. Illustrated. (In press.)

LANDRIN.—A TREATISE ON STEEL:
Comprising its Theory, Metallurgy, Properties, Practical Working, and Use. By M. H. C. LANDRIN, Jr., Civil Engineer. Translated from the French, with Notes, by A. A. FESQUET, Chemist and Engineer. With an Appendix on the Bessemer and the Martin Processes for Manufacturing Steel, from the Report of Abram S. Hewitt, United States Commissioner to the Universal Exposition, Paris, 1867. 12mo. $3 00

LANDRIN.—A TREATISE ON STEEL:
Comprising the Theory, Metallurgy, Practical Working, Properties, and Use. Translated from the French of H. C. LANDRIN, Jr., C. E. By A. A. FESQUET, Chemist and Engineer. Illustrated. 12mo. $3 00

LARKIN.—THE PRACTICAL BRASS AND IRON FOUNDER'S GUIDE:
A Concise Treatise on Brass Founding, Moulding, the Metals and their Alloys, etc.; to which are added Recent Improvements in the Manufacture of Iron, Steel by the Bessemer Process, etc. etc. By JAMES LARKIN, late Conductor of the Brass Foundry Department in Reany, Neafie & Co.'s Penn Works, Philadelphia. Fifth edition, revised, with Extensive additions. In one volume, 12mo. $2 25

LEAVITT.—FACTS ABOUT PEAT AS AN ARTICLE OF FUEL:
With Remarks upon its Origin and Composition, the Localities in which it is found, the Methods of Preparation and Manufacture, and the various Uses to which it is applicable; together with many other matters of Practical and Scientific Interest. To which is added a chapter on the Utilization of Coal Dust with Peat for the Production of an Excellent Fuel at Moderate Cost, especially adapted for Steam Service. By H. T. LEAVITT. Third edition. 12mo. . . . $1 75

LEROUX.—A PRACTICAL TREATISE ON THE MANUFACTURE OF WORSTEDS AND CARDED YARNS:
Translated from the French of CHARLES LEROUX, Mechanical Engineer, and Superintendent of a Spinning Mill. By Dr. H. PAINE, and A. A. FESQUET. Illustrated by 12 large plates. In one volume 8vo. $5 00

LESLIE (MISS).—COMPLETE COOKERY:
Directions for Cookery in its Various Branches. By MISS LESLIE. 60th edition. Thoroughly revised, with the addition of New Receipts. In 1 vol. 12mo., cloth . . $1 50

LESLIE (MISS). LADIES' HOUSE BOOK:
a Manual of Domestic Economy. 20th revised edition. 12mo., cloth $1 25

LESLIE (MISS).—TWO HUNDRED RECEIPTS IN FRENCH COOKERY.
12mo. 50

LIEBER.—ASSAYER'S GUIDE:
Or, Practical Directions to Assayers, Miners, and Smelters, for the Tests and Assays, by Heat and by Wet Processes, for the Ores of all the principal Metals, of Gold and Silver Coins and Alloys, and of Coal, etc. By OSCAR M. LIEBER. 12mo., cloth $1 25

LOVE.—THE ART OF DYEING, CLEANING, SCOURING, AND FINISHING:
On the most approved English and French methods; being Practical Instructions in Dyeing Silks, Woollens, and Cottons, Feathers, Chips, Straw, etc.; Scouring and Cleaning Bed and Window Curtains, Carpets, Rugs, etc.; French and English Cleaning, etc. By THOMAS LOVE. Second American Edition, to which are added General Instructions for the Use of Aniline Colors. 8vo. 5 00

MAIN AND BROWN.—QUESTIONS ON SUBJECTS CONNECTED WITH THE MARINE STEAM-ENGINE:
And Examination Papers; with Hints for their Solution. By Thomas J. Main, Professor of Mathematics, Royal Naval College, and Thomas Brown, Chief Engineer, R. N. 12mo., cloth $1 50

MAIN AND BROWN.—THE INDICATOR AND DYNAMOMETER:
With their Practical Applications to the Steam-Engine. By Thomas J. Main, M. A. F. R., Ass't Prof. Royal Naval College, Portsmouth, and Thomas Brown, Assoc. Inst. C. E., Chief Engineer, R. N., attached to the R. N. College. Illustrated. From the Fourth London Edition. 8vo. $1 50

MAIN AND BROWN.—THE MARINE STEAM-ENGINE.
By Thomas J. Main, F. R. Ass't S. Mathematical Professor at Royal Naval College, and Thomas Brown, Assoc. Inst. C. E. Chief Engineer, R. N. Attached to the Royal Naval College. Authors of "Questions Connected with the Marine Steam-Engine," and the "Indicator and Dynamometer." With numerous Illustrations. In one volume 8vo. $5 00

MARTIN.—SCREW-CUTTING TABLES, FOR THE USE OF MECHANICAL ENGINEERS:
Showing the Proper Arrangement of Wheels for Cutting the Threads of Screws of any required Pitch; with a Table for Making the Universal Gas-Pipe Thread and Taps. By W. A. Martin, Engineer. 8vo. 50

MILES—A PLAIN TREATISE ON HORSE-SHOEING.
With Illustrations. By William Miles, author of "The Horse's Foot" $1 00

MOLESWORTH.—POCKET-BOOK OF USEFUL FORMULÆ AND MEMORANDA FOR CIVIL AND MECHANICAL ENGINEERS.
By Guilford L. Molesworth, Member of the Institution of Civil Engineers, Chief Resident Engineer of the Ceylon Railway. Second American from the Tenth London Edition. In one volume, full bound in pocket-book form $2 00

MOORE.—THE INVENTOR'S GUIDE:
Patent Office and Patent Laws: or, a Guide to Inventors, and a Book of Reference for Judges, Lawyers, Magistrates, and others. By J G. Moore. 12mo., cloth $1 25

NAPIER.—A MANUAL OF ELECTRO-METALLURGY:
Including the Application of the Art to Manufacturing Processes. By James Napier. Fourth American, from the Fourth London edition, revised and enlarged. Illustrated by engravings. In one volume, 8vo. $2 00

NAPIER.—A SYSTEM OF CHEMISTRY APPLIED TO DYEING:
By JAMES NAPIER, F. C. S. A New and Thoroughly Revised Edition, completely brought up to the present state of the Science, including the Chemistry of Coal Tar Colors. By A. A. FESQUET, Chemist and Engineer. With an Appendix on Dyeing and Calico Printing, as shown at the Paris Universal Exposition of 1867, from the Reports of the International Jury, etc. Illustrated. In one volume 8vo., 400 pages $5 00

NEWBERY.—GLEANINGS FROM ORNAMENTAL ART OF EVERY STYLE;
Drawn from Examples in the British, South Kensington, Indian, Crystal Palace, and other Museums, the Exhibitions of 1851 and 1862, and the best English and Foreign works. In a series of one hundred exquisitely drawn Plates, containing many hundred examples. By ROBERT NEWBERY. 4to. $15 00

NICHOLSON.—A MANUAL OF THE ART OF BOOK-BINDING:
Containing full instructions in the different Branches of Forwarding, Gilding, and Finishing. Also, the Art of Marbling Book-edges and Paper. By JAMES B. NICHOLSON. Illustrated. 12mo. cloth $2 25

NORRIS.—A HAND-BOOK FOR LOCOMOTIVE ENGINEERS AND MACHINISTS:
Comprising the Proportions and Calculations for Constructing Locomotives; Manner of Setting Valves; Tables of Squares, Cubes, Areas, etc. etc. By SEPTIMUS NORRIS, Civil and Mechanical Engineer. New edition. Illustrated, 12mo., cloth
$2 00

NYSTROM.—ON TECHNOLOGICAL EDUCATION AND THE CONSTRUCTION OF SHIPS AND SCREW PROPELLERS:
For Naval and Marine Engineers. By JOHN W. NYSTROM, late Acting Chief Engineer U. S. N. Second edition, revised with additional matter. Illustrated by seven engravings. 12mo.
$2 50

O'NEILL.—A DICTIONARY OF DYEING AND CALICO PRINTING:
Containing a brief account of all the Substances and Processes in use in the Art of Dyeing and Printing Textile Fabrics: with Practical Receipts and Scientific Information. By CHARLES O'NEILL, Analytical Chemist; Fellow of the Chemical Society of London; Member of the Literary and Philosophical Society of Manchester; Author of "Chemistry of Calico Printing and Dyeing." To which is added An Essay on Coal Tar Colors and their Application to

Dyeing and Calico Printing. By A. A. FESQUET, Chemist and Engineer. With an Appendix on Dyeing and Calico Printing, as shown at the Exposition of 1867, from the Reports of the International Jury, etc. In one volume 8vo., 491 pages . . $6 00

OSBORN.—THE METALLURGY OF IRON AND STEEL:
Theoretical and Practical : In all its Branches ; With Special Reference to American Materials and Processes. By H. S. OSBORN, LL. D., Professor of Mining and Metallurgy in Lafayette College, Easton, Pa. Illustrated by 230 Engravings on Wood, and 6 Folding Plates. 8vo., 972 pages $10 00

OSBORN.—AMERICAN MINES AND MINING :
Theoretically and Practically Considered. By Prof. H. S. OSBORN, Illustrated by numerous engravings. 8vo. (*In preparation.*)

PAINTER, GILDER, AND VARNISHER'S COMPANION:
Containing Rules and Regulations in everything relating to the Arts of Painting, Gilding, Varnishing, and Glass Staining, with numerous useful and valuable Receipts; Tests for the Detection of Adulterations in Oils and Colors, and a statement of the Diseases and Accidents to which Painters, Gilders, and Varnishers are particularly liable, with the simplest methods of Prevention and Remedy. With Directions for Graining, Marbling, Sign Writing, and Gilding on Glass. To which are added COMPLETE INSTRUCTIONS FOR COACH PAINTING AND VARNISHING. 12mo., cloth, $1 50

PALLETT.—THE MILLER'S, MILLWRIGHT'S, AND ENGINEER'S GUIDE.
By HENRY PALLETT. Illustrated. In one vol. 12mo. . $3 00

PERKINS.—GAS AND VENTILATION.
Practical Treatise on Gas and Ventilation. With Special Relation to Illuminating, Heating, and Cooking by Gas. Including Scientific Helps to Engineer-students and others. With illustrated Diagrams. By E. E. PERKINS. 12mo., cloth . . . $1 25

PERKINS AND STOWE.—A NEW GUIDE TO THE SHEET-IRON AND BOILER PLATE ROLLER:
Containing a Series of Tables showing the Weight of Slabs and Piles to Produce Boiler Plates, and of the Weight of Piles and the Sizes of Bars to Produce Sheet-Iron; the Thickness of the Bar Gauge in Decimals; the Weight per foot, and the Thickness on the Bar or Wire Gauge of the fractional parts of an inch; the Weight per sheet, and the Thickness on the Wire Gauge of Sheet-Iron of various dimensions to weigh 112 lbs. per bundle; and the conversion of Short Weight into Long Weight, and Long Weight into Short. Estimated and collected by G. H. PERKINS and J. G. STOWE $2 50

PHILLIPS AND DARLINGTON.—RECORDS OF MINING AND METALLURGY:
Or, Facts and Memoranda for the use of the Mine Agent and Smelter. By J. ARTHUR PHILLIPS, Mining Engineer, Graduate of the Imperial School of Mines, France, etc., and JOHN DARLINGTON. Illustrated by numerous engravings. In one vol. 12mo. . $2 00

PRADAL, MALEPEYRE, AND DUSSAUCE.—A COMPLETE TREATISE ON PERFUMERY:
Containing notices of the Raw Material used in the Art, and the Best Formulæ. According to the most approved Methods followed in France, England, and the United States. By M. P. PRADAL, Perfumer-Chemist, and M. F. MALEPEYRE. Translated from the French, with extensive additions, by Prof. H. DUSSAUCE. 8vo. $10

PROTEAUX.—PRACTICAL GUIDE FOR THE MANUFACTURE OF PAPER AND BOARDS.
By A. PROTEAUX, Civil Engineer, and Graduate of the School of Arts and Manufactures, Director of Thiers's Paper Mill, 'Puy-de-Dôme. With additions, by L. S. LE NORMAND. Translated from the French, with Notes, by HORATIO PAINE, A. B., M. D. To which is added a Chapter on the Manufacture of Paper from Wood in the United States, by HENRY T. BROWN, of the "American Artisan." Illustrated by six plates, containing Drawings of Raw Materials, Machinery, Plans of Paper-Mills, etc. etc. 8vo. $5 00

REGNAULT.—ELEMENTS OF CHEMISTRY.
By M. V. REGNAULT. Translated from the French by T. FORREST BETTON, M. D., and edited, with notes, by JAMES C. BOOTH, Melter and Refiner U. S. Mint, and WM. L. FABER, Metallurgist and Mining Engineer. Illustrated by nearly 700 wood engravings. Comprising nearly 1500 pages. In two vols. 8vo., cloth $10 00

REID.—A PRACTICAL TREATISE ON THE MANUFACTURE OF PORTLAND CEMENT:
By HENRY REID, C. E. To which is added a Translation of M. A. Lipowitz's Work, describing a new method adopted in Germany of Manufacturing that Cement. By W. F. REID. Illustrated by plates and wood engravings. 8vo. $7 00

RIFFAULT, VERGNAUD, AND TOUSSAINT.—A PRACTICAL TREATISE ON THE MANUFACTURE OF COLORS FOR PAINTING:
Containing the best Formulæ and the Processes the Newest and in most General Use. By MM. RIFFAULT, VERGNAUD, and TOUSSAINT. Revised and Edited by M. F. MALEPEYRE and Dr. EMIL WINCKLER. Illustrated by Engravings. In one vol. 8vo. (*In preparation.*)

RIFFAULT, VERGNAUD, AND TOUSSAINT.—A PRACTICAL TREATISE ON THE MANUFACTURE OF VARNISHES:
By MM. RIFFAULT, VERGNAUD, and TOUSSAINT. Revised and Edited by M. F. MALEPEYRE and Dr. EMIL. WINCKLER. Illustrated. In one vol. 8vo. (*In preparation.*)

SHUNK.—A PRACTICAL TREATISE ON RAILWAY CURVES AND LOCATION, FOR YOUNG ENGINEERS.
By WM. F. SHUNK, Civil Engineer. 12mo., tucks . . $2 00

SMEATON.—BUILDER'S POCKET COMPANION:
Containing the Elements of Building, Surveying, and Architecture; with Practical Rules and Instructions connected with the subject. By A. C. SMEATON, Civil Engineer, etc. In one volume, 12mo. $1 50

SMITH.—THE DYER'S INSTRUCTOR:
Comprising Practical Instructions in the Art of Dyeing Silk, Cotton, Wool, and Worsted, and Woollen Goods: containing nearly 800 Receipts. To which is added a Treatise on the Art of Padding; and the Printing of Silk Warps, Skeins, and Handkerchiefs, and the various Mordants and Colors for the different styles of such work. By DAVID SMITH, Pattern Dyer, 12mo., cloth
$3 00

SMITH.—THE PRACTICAL DYER'S GUIDE:
Comprising Practical Instructions in the Dyeing of Shot Cobourgs, Silk Striped Orleans, Colored Orleans from Black Warps, ditto from White Warps, Colored Cobourgs from White Warps, Merinos, Yarns, Woollen Cloths, etc. Containing nearly 300 Receipts, to most of which a Dyed Pattern is annexed. Also, a Treatise on the Art of Padding. By DAVID SMITH. In one vol. 8vo. $25 00

SHAW.—CIVIL ARCHITECTURE:
Being a Complete Theoretical and Practical System of Building, containing the Fundamental Principles of the Art. By EDWARD SHAW, Architect. To which is added a Treatise on Gothic Architecture, &c. By THOMAS W. SILLOWAY and GEORGE M. HARDING, Architects. The whole illustrated by 102 quarto plates finely engraved on copper. Eleventh Edition. 4to. Cloth. $10 00

SLOAN.—AMERICAN HOUSES:
A variety of Original Designs for Rural Buildings. Illustrated by 26 colored Engravings, with Descriptive References. By SAMUEL SLOAN, Architect, author of the "Model Architect," etc. etc. 8vo.
$2 50

SMITH.—PARKS AND PLEASURE GROUNDS:
Or, Practical Notes on Country Residences, Villas, Public Parks, and Gardens. By CHARLES H. J. SMITH, Landscape Gardener and Garden Architect, etc. etc. 12mo. $2 25

STOKES.—CABINET-MAKER'S AND UPHOLSTERER'S COMPANION:
Comprising the Rudiments and Principles of Cabinet-making and Upholstery, with Familiar Instructions, Illustrated by Examples for attaining a Proficiency in the Art of Drawing, as applicable to Cabinet-work; The Processes of Veneering, Inlaying, and Buhl-work; the Art of Dyeing and Staining Wood, Bone, Tortoise Shell, etc. Directions for Lackering, Japanning, and Varnishing; to make French Polish; to prepare the Best Glues, Cements, and Compositions, and a number of Receipts, particularly for workmen generally. By J. STOKES. In one vol. 12mo. With illustrations
$1 25

STRENGTH AND OTHER PROPERTIES OF METALS.
Reports of Experiments on the Strength and other Properties of Metals for Cannon. With a Description of the Machines for Testing Metals, and of the Classification of Cannon in service. By Officers of the Ordnance Department U. S. Army. By authority of the Secretary of War. Illustrated by 25 large steel plates. In 1 vol. quarto $10 00

TABLES SHOWING THE WEIGHT OF ROUND, SQUARE, AND FLAT BAR IRON, STEEL, ETC.
By Measurement. Cloth 63

TAYLOR.—STATISTICS OF COAL:
Including Mineral Bituminous Substances employed in Arts and Manufactures; with their Geographical, Geological, and Commercial Distribution and amount of Production and Consumption on the American Continent. With Incidental Statistics of the Iron Manufacture. By R. C. TAYLOR. Second edition, revised by S. S. HALDEMAN. Illustrated by five Maps and many wood engravings. 8vo., cloth $6 00

TEMPLETON.—THE PRACTICAL EXAMINATOR ON STEAM AND THE STEAM-ENGINE:
With Instructive References relative thereto, for the Use of Engineers, Students, and others. By WM. TEMPLETON, Engineer 12mo.
$1 25

THOMAS.—THE MODERN PRACTICE OF PHOTOGRAPHY.
By R. W. THOMAS, F. C. S. 8vo., cloth 75

THOMSON.—FREIGHT CHARGES CALCULATOR.
By Andrew Thomson, Freight Agent $1 25

TURNING: SPECIMENS OF FANCY TURNING EXECUTED ON THE HAND OR FOOT LATHE:
With Geometric, Oval, and Eccentric Chucks, and Elliptical Cutting Frame. By an Amateur. Illustrated by 30 exquisite Photographs. 4to. $3 00

TURNER'S (THE) COMPANION:
Containing Instructions in Concentric, Elliptic, and Eccentric Turning; also various Plates of Chucks, Tools, and Instruments; and Directions for using the Eccentric Cutter, Drill, Vertical Cutter, and Circular Rest; with Patterns and Instructions for working them. A new edition in 1 vol. 12mo. $1 50

URBIN—BRULL.—A PRACTICAL GUIDE FOR PUDDLING IRON AND STEEL.
By Ed. Urbin, Engineer of Arts and Manufactures. A Prize Essay read before the Association of Engineers, Graduate of the School of Mines, of Liege, Belgium, at the Meeting of 1865-6. To which is added a Comparison of the Resisting Properties of Iron and Steel. By A. Brull. Translated from the French by A. A. Fesquet, Chemist and Engineer. In one volume, 8vo.
$1 00

WARN.—THE SHEET METAL WORKER'S INSTRUCTOR, FOR ZINC, SHEET-IRON, COPPER AND TIN PLATE WORKERS, &c.
By Reuben Henry Warn, Practical Tin Plate Worker. Illustrated by 32 plates. 8vo. (*In press, nearly ready.*)

WATSON.—A MANUAL OF THE HAND-LATHE.
By Egbert P. Watson, Late of the "Scientific American," Author of "Modern Practice of American Machinists and Engineers," In one volume, 12mo. $1 50

WATSON.—THE MODERN PRACTICE OF AMERICAN MACHINISTS AND ENGINEERS:
Including the Construction, Application, and Use of Drills, Lathe Tools, Cutters for Boring Cylinders, and Hollow Work Generally, with the most Economical Speed of the same, the Results verified by Actual Practice at the Lathe, the Vice, and on the Floor. Together with Workshop management, Economy of Manufacture, the Steam-Engine, Boilers, Gears, Belting, etc. etc. By Egbert P. Watson, late of the "Scientific American." Illustrated by eighty-six engravings. 12mo. $2 50

WATSON.—THE THEORY AND PRACTICE OF THE ART OF WEAVING BY HAND AND POWER:
With Calculations and Tables for the use of those connected with the Trade. By JOHN WATSON, Manufacturer and Practical Machine Maker. Illustrated by large drawings of the best Power-Looms. 8vo. $10 00

WEATHERLY.—TREATISE ON THE ART OF BOILING SUGAR, CRYSTALLIZING, LOZENGE-MAKING, COMFITS, GUM GOODS,
And other processes for Confectionery, &c. In which are explained, in an easy and familiar manner, the various Methods of Manufacturing every description of Raw and Refined Sugar Goods, as sold by Confectioners and others . . . $2 00

WILL.—TABLES FOR QUALITATIVE CHEMICAL ANALYSIS.
By Prof. HEINRICH WILL, of Giessen, Germany. Seventh edition. Translated by CHARLES F. HIMES, Ph. D., Professor of Natural Science, Dickinson College, Carlisle, Pa. . . $1 25

WILLIAMS.—ON HEAT AND STEAM:
Embracing New Views of Vaporization, Condensation, and Expansion. By CHARLES WYE WILLIAMS, A. I. C. E. Illustrated. 8vo.
$3 50

WOHLER.—A PRACTICAL TREATISE ON ANALYTICAL CHEMISTRY.
By F. Wöhler. With additions by GRANDEAU and TROOST. Edited by H. B. NASON, Professor of Chemistry, Rensselaer Institute, Troy, N. Y. With numerous Illustrations. (*In press.*)

WORSSAM.—ON MECHANICAL SAWS:
From the Transactions of the Society of Engineers, 1867. By S. W. WORSSAM, Jr. Illustrated by 18 large folding plates. 8vo.
$5 00

www.ingramcontent.com/pod-product-compliance
Lightning Source LLC
Chambersburg PA
CBHW020108170426
43199CB00009B/447